Nuclear Engineering Division 2015

Core Programming Area at the 2015 AIChE Annual Meeting

Salt Lake City, Utah, USA
8-13 November 2015

ISBN: 978-1-5108-1868-2

Printed from e-media with permission by:

Curran Associates, Inc.
57 Morehouse Lane
Red Hook, NY 12571

Some format issues inherent in the e-media version may also appear in this print version.

Printed by Curran Associates, Inc. (2016)

For permission requests, please contact AIChE
at the address below.

AIChE
120 Wall Street, FL 23
New York, NY 10005-4020

Phone: (800) 242-4363
Fax: (203) 775-5177

www.aiche.org

Additional copies of this publication are available from:

Curran Associates, Inc.
57 Morehouse Lane
Red Hook, NY 12571 USA
Phone: 845-758-0400
Fax: 845-758-2634
Email: curran@proceedings.com
Web: www.proceedings.com

TABLE OF CONTENTS

409853 Metal-Ferrocyanide Functionalized Magnetic Adsorbents for the Removal of Radioactive Cesium in Contaminated Water

Abstract:

The nuclear accident at the Fukushima Daiichi nuclear power station in 2011 released a huge quantity of radioactive contaminants into the environment [1]. Among these, ^{137}Cs is the most problematic contaminant due to its long half-life (30.2 years), high solubility in water, and high-energy gamma ray emissions. Various types of Cs adsorbents such as zeolites, calixarene, aluminum molybdophosphate, and metal ferrocyanide have been investigated. [2]. However, they cannot be easily used to remove Cs in an open environment because there is no easy way to collect the used adsorbents after applying them into the environment. Magnetic adsorbents have attracted significant attention for environmental remediation application, as they display a high surface area and provide unique magnetic separation property.

In this study, metal (cupper or iron) ferrocyanide-functionalized magnetic adsorbents were fabricated for the highly efficient removal of radioactive cesium from contaminated water. The hydrophilic polymer such as polyethyleneimine, polyvinylpyrrolidone were coated on the surface of magnetic nanocluster to stabilize the magnetic adsorbent in water and used as coating materials to immobilize the copper and ferrocyanide on the surface of magnetic nanoclusters. Furthermore, Iron ferrocyanide was directly coated onto the surface of magnetic nanoparticles without template such as hydrophilic polymer for the enhanced magnetic property and Cs removal performance. The resulting metal-ferrocyanide functionalized magnetic adsorbents having supeparamagnetic property could be easily separated from water using an external magnet, and also showed excellent removal efficiency of radioactive cesium from contaminated water due to the ion exchange capacity of the metal-ferrocyanide.

References

1. M. Manolopoulou, E. Vagena, S. Stoulos, A. Ioannidou, C. Papastefanou, J. Environ. Radioact. 102, 796 (2011).

2. P. A. Haas, Sep. Sci. Technol. 28, 2479 (1993).

411131 Confined Swelling Pressure Tests for Spherical Resorcinol-Formaldehyde Resin

Abstract:

Confined Swelling Pressure Tests for Spherical Resorcinol-Formaldehyde Resin

Joseph Walker, Paul Taylor, and Trent Walker

Oak Ridge National Laboratory

An ion exchange process using spherical resorcinol-formaldehyde (sRF) resin from Microbeads AS (Skedsmokorset, Norway) is the baseline process for removing cesium from the approximately 53 million gallons of highly radioactive nuclear waste currently stored in underground tanks at the Department of Energy (DOE) Hanford site. Since the selected sRF resin is known to swell during processing as it is converted from a hydrogen form to a sodium form, there is a concern about the pressure the resin could exert if it were to become trapped in process piping.

A test system (see Fig. 1) was fabricated and operated at Oak Ridge National Laboratory to measure the swelling pressure of various batches of sRF resin. The test vessel consists of a 6 inch long section of 2 inch, schedule 80, 304L stainless steel pipe welded to 2 inch, class 600 pipe flanges at the top and bottom. With the flanges, the total depth of the resin bed within the test vessel is 6.5 inches. The inlet and outlet filter assemblies, which are modified Swagelok ½-in. stainless inline filters, are welded to the sides of the vessel. Pressure elements/transmitters are bolted to the top and bottom flanges of the test vessel to monitor the pressures during testing. The top pressure assembly is a 0–750 psi, WIKA general industrial transmitter (WIKA Instrument, LP, Lawrenceville, GA) with a flush diaphragm seal, and the bottom pressure assembly is a 0–1500 psi assembly of the same design.

Fig. 1. Photograph of the test system before instruments were connected to the data acquisition system.

Six batches of the sRF resin were tested:

1. Blue drum (batch # 5J-370/686)
2. Black drum (batch # BSC-3380-3-0200)
3. Finncont-1 (batch # BSC-3380-2P-0201)
4. Finncont-2 (batch # 6C-370/745)
5. Finncont-3 (batch # 0L-370-1312)
6. Microbeads (batch # 1f-370-1392)

For each test, the test vessel was filled to the top flange (see Fig. 2) with preconditioned, hydrogen-form sRF resin in water, and then the top flange/pressure transducer was connected, air was vented from the lines, and dilute sodium hydroxide was pumped through the resin to convert it to the hydroxide form. The pressures at the top and bottom of the test vessel were measured as the resin swelled inside the test vessel. When the pressures had stabilized, the flow of sodium hydroxide was stopped so that the static pressures inside the test vessel could be measured. A small amount of deionized water was then pumped through the test vessel to displace the sodium hydroxide, and the top flange was removed, allowing the resin to expand (see Fig. 3).

Fig. 2. H-form sRF resin loaded in test vessel **Fig. 3. sRF resin expansion after top flange was removed**

One run each was made with the first five resin batches listed above, and five runs were made with the sixth batch. After each test, samples of resin were collected from different locations within the column, titrated to determine the extent of conversion to the sodium form, and examined microscopically to look for bead breakage and changes in the bead sizes and shapes.

426160 Analysis of Mobile Helium Cluster Reactions Near Surfaces and Grain Boundaries of Plasma-Exposed Tungsten

Abstract:

The evolution of the surface morphology and the near-surface structure of plasma facing components (PFCs) in nuclear fusion reactors is impacted significantly by the implantation of helium (He) atoms. Tungsten (W) is an important PFC material due to its thermomechanical properties. In tungsten, such interstitial He atoms are very mobile and aggregate to form clusters of different sizes. The smallest of these helium clusters, containing n helium atoms with n = 1-7, also are mobile and their diffusional transport mediates the evolution of surface morphology and the sub-surface helium gas bubble structure and dynamics.

In this presentation, we report the results of a systematic atomic-scale analysis of the reactions of small mobile helium clusters near low-Miller-index tungsten surfaces, aiming at a fundamental understanding of the near-surface dynamics of helium-carrying species in plasma-exposed tungsten. These small mobile helium clusters are attracted to the surface and migrate to the surface by Fickian diffusion and drift due to the thermodynamic driving force for surface segregation. As the clusters migrate toward the surface, trap mutation (TM) and cluster dissociation reactions are activated at rates higher than in the bulk. TM produces W adatoms and immobile complexes of helium clusters surrounding W vacancies located within the lattice planes at a short distance from the surface. These reactions are identified and characterized in detail based on analysis of a large number of molecular-dynamics (MD) trajectories for each such mobile cluster near W(100), W(110), and W(111) surfaces. TM is found to be the dominant cluster reaction for all cluster and surface combinations, except for the clusters with n = 4 and n = 5 near W(100) where cluster partial dissociation following TM dominates. We find that there exists a critical cluster size, n = 4 near W(100) and W(111) and n = 5 near W(110), beyond which formation of multiple W adatoms and vacancies in TM reactions is observed. The identified cluster reactions are responsible for important structural, morphological, and compositional features in plasma-exposed tungsten, including surface adatom populations, near-surface immobile helium-vacancy complexes, and retained helium content, which are expected to influence the amount of hydrogen re-cycling and tritium retention in fusion tokamaks.

We have also carried out a systematic atomic-scale analysis of reactions of small mobile helium clusters near a symmetric tilt grain boundary (GB) based on MD simulations. We found that TM reactions are dominant for clusters with n = 4-7, and only a single vacancy is formed in each of the TM reactions identified regardless of cluster size n. The displaced W atom generated from the TM reaction near the GB forms an extended W interstitial configuration on the GB. We also found that, for n = 1-4, the helium clusters (including single He interstitial atoms) that segregate on the GB are rendered practically immobile. However, the mobility of the extended W interstitial configuration on the GB depends on the exact location where the TM reactions take place, with the W interstitial being highly mobile if the TM reaction occurs on the GB. Although this study is limited to a simple, prototypical model GB structure, the insights obtained are very useful for understanding surface morphology and helium retention in plasma-exposed polycrystalline tungsten.

428651 High Operating Temperature Transfer and Storage (HOTTS) System for Concentrated Solar Power Generation and Industrial Heat Integration

Abstract:

Heat transfer and storage technologies for industrial processes, such as concentrated solar power and metals production, are limited by the high temperature stability of their constituent materials. We are developing a High Operating Temperature Transfer and Storage (HOTTS) process, which will enable high quality heat integration at temperatures up to and exceeding 1,100°C. The HOTTS process integrates two key technologies: a high-temperature, material of construction (MOC) and a novel particle-based heat transfer fluid (pHTF). The MOC is composed of Mo-Si-B composites, which exhibit excellent oxidative and erosive resistance. The pHTF flows freely, can be conveyed by air, and resists sintering in overnight storage conditions.

In this study, we report on approaches to apply the MOC as a protective coating for molybdenum substrates, including a solution-based technique. We compare the coating coverage, morphology, and substrate adhesion to the oxidative stability of the composite at temperatures above 1,100°C in air. This composite material offers the potential to combine the excellent high temperature mechanical properties of Mo (e.g., creep strength) with the oxidative protection of Mo-Si-B. We also discuss the development of a pilot-scale system for evaluating the heat transfer performance of the pHTF. Heat-transfer coefficients are determined for mass fluxes above 100 kg/m^2-s in moving- and fluidized-bed-type configurations. Engineering approaches to integrate the MOC into the HOTTS process, as well as heat integration strategies into industrial processes will also be discussed.

Discrete Element and Continuum Heat Transfer Simulations of a Solar Receiver That Uses Solid Particles As a Heat Transfer Fluid

2015 AIChE Annual Meeting

November 9, 2015

Dr. Aaron Morris	University of Colorado
Prof. Christine Hrenya	University of Colorado (PI)
Dr. Zhiwen Ma	NREL (co-PI)
Dr. Sreekanth Pannala	ORNL (co-PI)
Dr. Tom O'Brien	retired NETL (co-PI)

Powered by SunShot U.S. Department of Energy

Concentrating Solar Power (CSP)

Powered by SunShot U.S. Department of Energy

Power Tower CSP system

- Heliostats focus energy on a central tower, warming a heat transfer fluid that can in turn power a turbine
- Aim to improve upon performance by using *solid particles* as a heat transfer fluid and storage media

2/23

Challenges for CSP technology

How different heat transfer fluids impact design

- Operate at high temperatures; $T\uparrow$ *Efficiency* $\uparrow$
- Limit on/off cycling; Day/night and cloud cover
  - Thermal storage (Run 24hr/day)

3/23

Challenges for CSP technology

How different heat transfer fluids impact design

- Operate at high temperatures; $T\uparrow$ *Efficiency* $\uparrow$
- Limit on/off cycling; Day/night and cloud cover
  - Thermal storage (Run 24hr/day)

Candidate Fluids	High Temp?	Thermal Storage?
Steam	Yes	No *High pressure overhead*
Molten Salts	No can be unstable	Yes *Excellent chemical storage*

4/23

Challenges for CSP technology

How different heat transfer fluids impact design

- Operate at high temperatures; *T* ↑ *Efficiency* ↑

- Limit on/off cycling; Day/night and cloud cover
  - Thermal storage (Run 24hr/day)

Candidate Fluids	High *T*?	Thermal Storage?
Steam	Yes	No *High storage costs*
Molten Salts	No *can be unstable*	Yes
Sand	Yes	Yes

5/23

NREL SunShot CSP Award (PI: Zhiwen Ma)

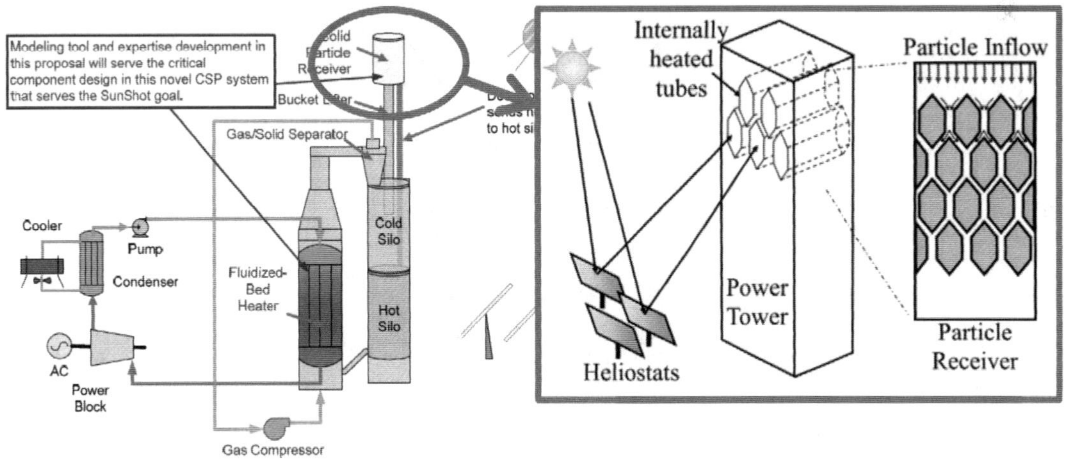

Overall Goal: *<u>Fundamental, continuum</u> modeling tool that can be used for design of particle receiver: understanding and prediction of <u>heat transfer in solids flows</u>*

Why fundamental? *Counterintuitive/non-monotonic trends, correlations unreliable for extrapolation*

Why continuum model? *Discrete-particle models computationally infeasible*

6/23

Model Development for Solar Collector

Overall Approach: Use DEM simulations as "ideal experiment" to test continuum theory

- DEM: instantaneous, local, non-intrusive measurements of concentration, velocity, and temperature
- DEM: can turn various mechanisms on/off to assess relative contributions

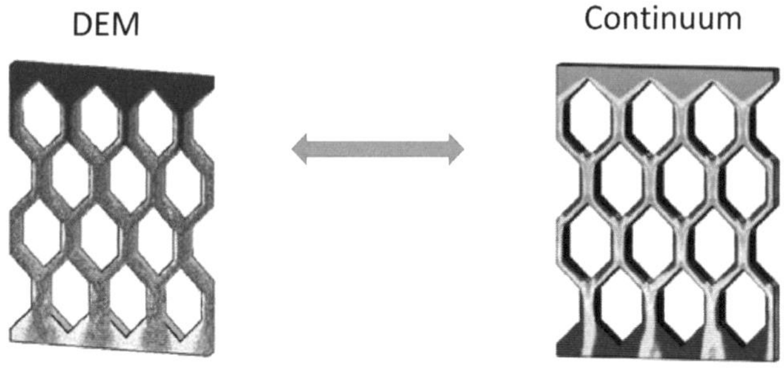

Computational Tool: MFIX (Multiphase Flow with Interface eXchanges)

Heat transfer mechanisms

Conduction

- *Particle-wall* and particle-fluid-wall

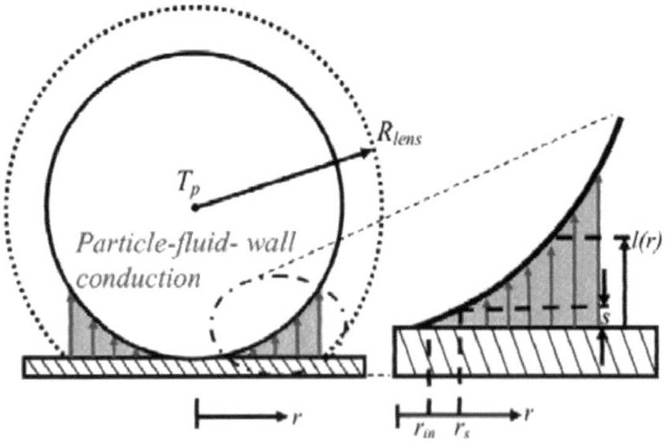

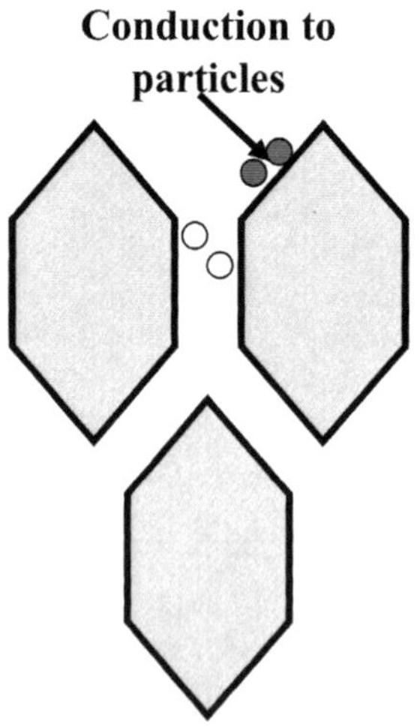

Heat transfer mechanisms

Powered by SunShot U.S. Department of Energy

Conduction

- *Particle-wall* and particle-fluid-wall

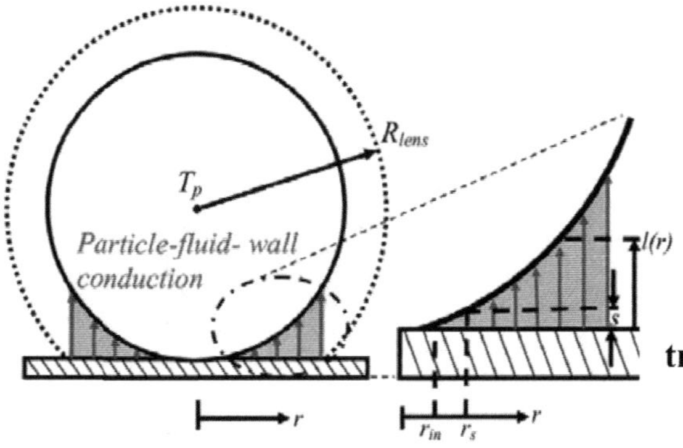

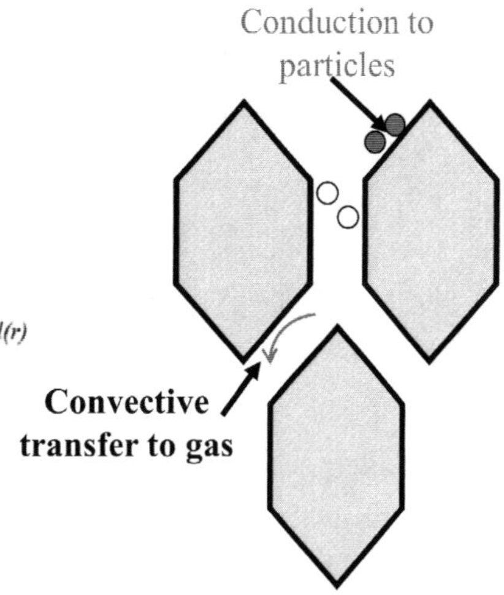

Convection

- Entrained gas warmed heated and then heats particles

Heat transfer to flowing particles

Powered by SunShot U.S. Department of Energy

Conductive heat transfer: Particle vs. molecular fluids

Thermal contact “resistance” for particle flows

- *Thermal slip* for particle flows – molecular fluids often *no slip*

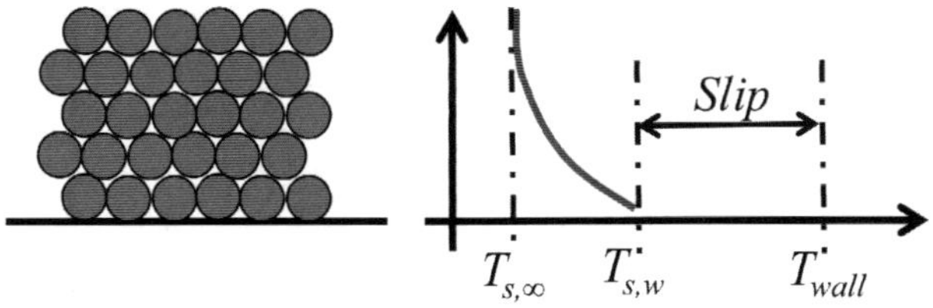

Heat transfer to flowing particles

Conductive heat transfer: Particle vs. molecular fluids

Thermal contact "resistance" for particle flows

- *Thermal slip* for particle flows – molecular fluids often *no slip*

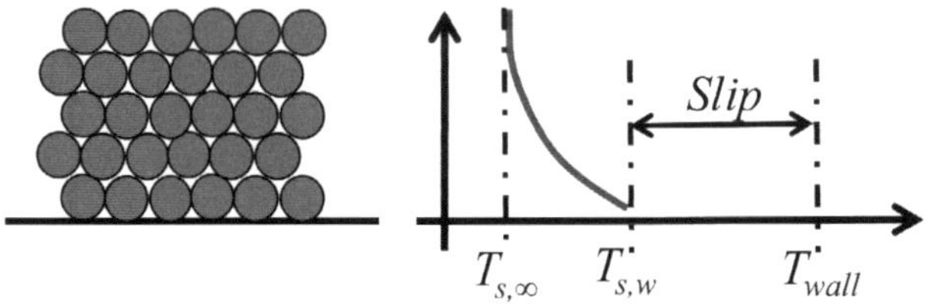

- Contact resistance depends on how particles contact wall
  - Solids concentration near wall
  - Particle size / morphology

11/23

Heat transfer to flowing particles

Conductive heat transfer: Particle vs. molecular fluids

Thermal contact "resistance" for particle flows

- *Thermal slip* for particle flows – molecular fluids often *no slip*

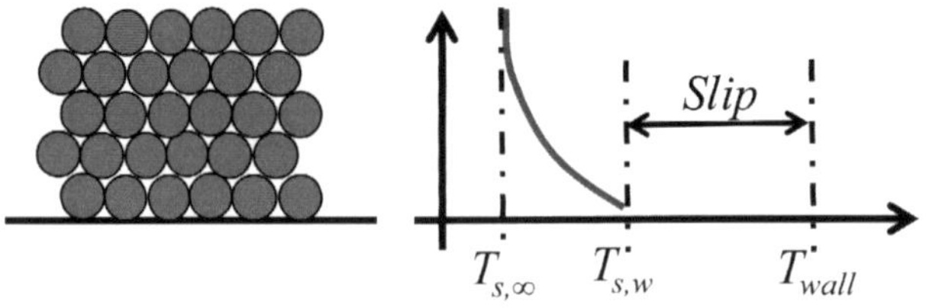

- Contact resistance depends on how particles contact wall
  - Solids concentration near wall
  - Particle size / morphology

- Often modeled using Nusselt number correlation
  - **Existing models are empirical or analytically derived for packed beds. Found inaccurate for dynamic systems**

12/23

New continuum heat transfer model

Derive a continuum (large-scale) closure from particle-scale fundamental model

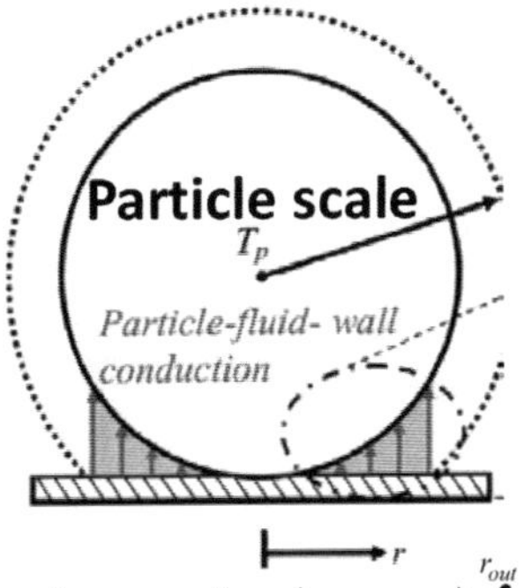

$$\dot{Q}_{pfw} = H\left(T_w - T_p\right) = \left(T_w - T_p\right)\int_{r_{in}}^{r_{out}} \frac{2\pi\kappa_g r}{\max(l,s)}\, dr$$

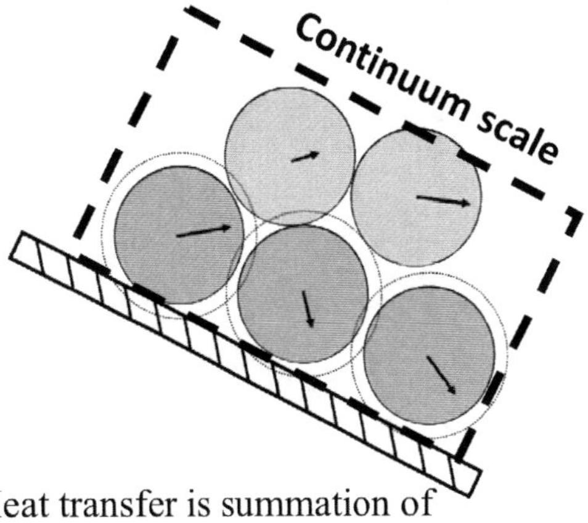

Heat transfer is summation of contributions from all particles

$$\dot{Q}_{wall} = \sum_{i=1}^{N_p} \dot{Q}_{pfw,i}$$

Depends on proximity to wall (unknown in continuum framework)

13/23

New continuum heat transfer model

Non-dimensionalize and *integrate* over all particles near wall

- Location represented by "particle-wall distribution function"

$$\mathrm{Nu} = \frac{hD_p}{\kappa_g} = \frac{1}{2}\left(\frac{\varepsilon_s}{\varepsilon_{s,\max}}\right)^{2/3} \frac{\int_0^{1-\hat{R}_{lens}} \chi(\hat{\delta})\hat{H}(\hat{\delta})\,d\hat{\delta}}{\int_0^{1-\hat{R}_{lens}} \chi(\hat{\delta})\,d\hat{\delta}}$$

$$q''_{pfw} = h(T_w - T_s)$$

Separation from wall

"classic"
radial distribution function $g(r)$

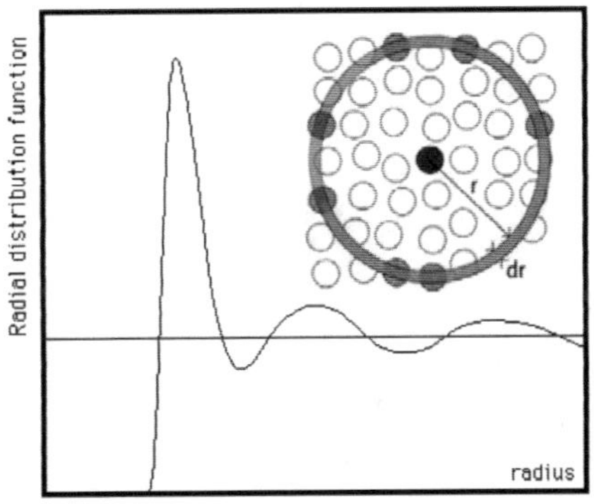

new "particle-wall"
distribution function $\chi(\delta)$
Measured from DEM

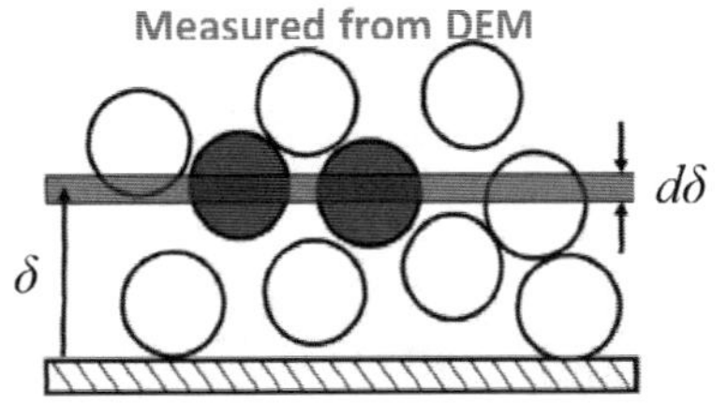

[1]Morris, A. B., Pannala, S., Ma, Z., and Hrenya, C. M., "A conductive heat transfer model for particle flows over immersed surfaces," *International Journal of Heat and Mass Transfer*, vol. 89, 2015, pp. 1277-1289.

14/23

Local Heat Transfer Coefficient: New Model vs. Previous

New model tested for particle flow down top surface of hexagon

- Not tested for full receiver

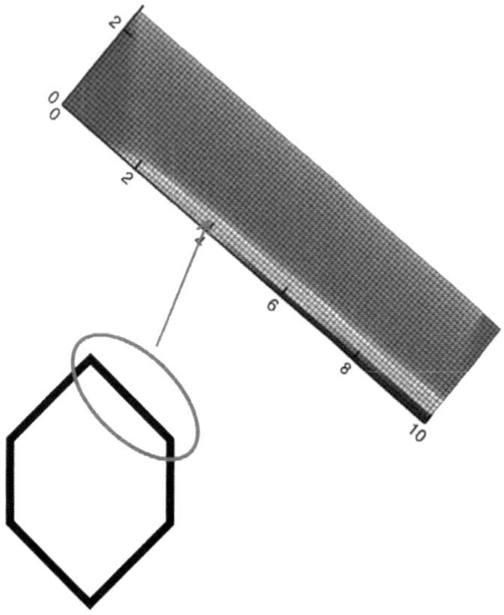

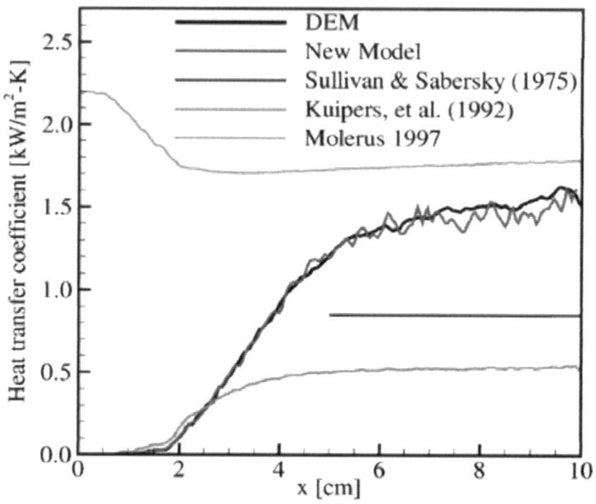

15/23

DEM Simulations of hexagonal tube array

Base case

Particle Diameter	300 μm
e (rest. coeff)	0.9
Solids Mass Flux	258.2 kg/m²-s
Inlet Temperature	300 K
Wall Temperature	600 K
X-spacing	5.04 cm
Y-spacing	4.98 cm
Hexagon Width	3.78 cm
Hexagon Height	6.42 cm
Apex Angle	89°

~ 6 million particles

Also varied…

- Tube spacing
- Apex angle
- Flow rate
- Particle size

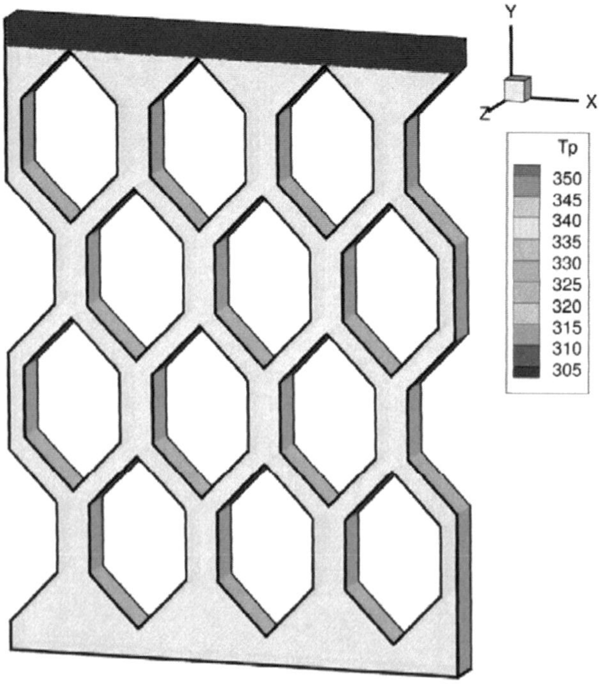

16/23

DEM Simulations of hexagonal tube array

Role of Apex Angle

Apex Angle	Average h [W/m^2-K]
89°	487.41
75°	371.42
60°	344.14

$$\bar{h} = \frac{\dot{Q}_{tot}}{A_{hex,tot}\left(T_{wall} - T_{s,inlet}\right)}$$

- Heat transfer decreases as apex steepens

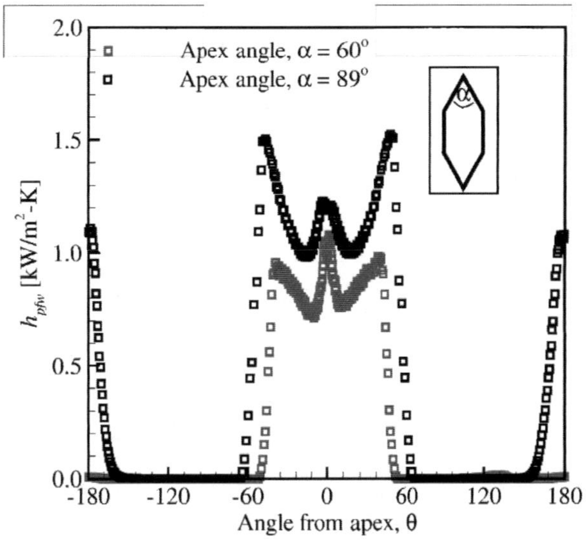

DEM Simulations of hexagonal tube array

Role of Apex Angle

Apex Angle	Average h [W/m^2-K]
89°	487.41
75°	371.42
60°	344.14

$$\bar{h} = \frac{\dot{Q}_{tot}}{A_{hex,tot}\left(T_{wall} - T_{s,inlet}\right)}$$

- Heat transfer decreases as apex steepens and as velocity increases
- *Opposite* trend for molecular fluids

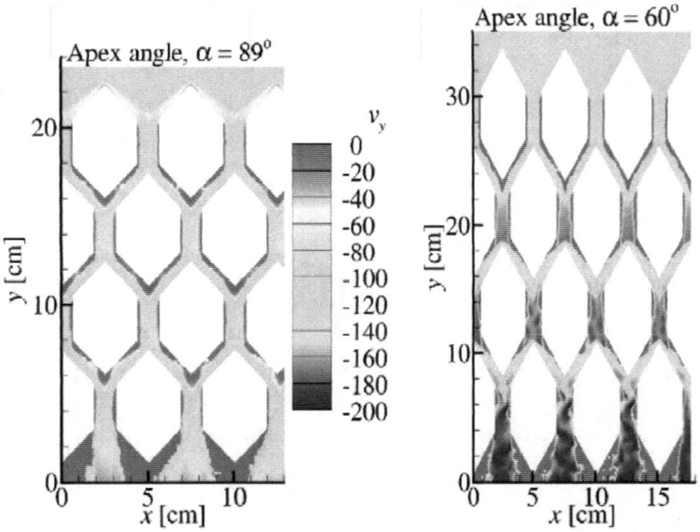

DEM Simulations of hexagonal tube array

Role of Particle Size

Particle size	Average h [W/m²-K]
300 micron	487.41
200 micron	780.34
100 micron	997.23**

$$\overline{h} = \frac{\dot{Q}_{tot}}{A_{hex,tot}\left(T_{wall} - T_{s,inlet}\right)}$$

**** Only two rows of hexagons simulated**

- Slightly stronger instabilities for small particles

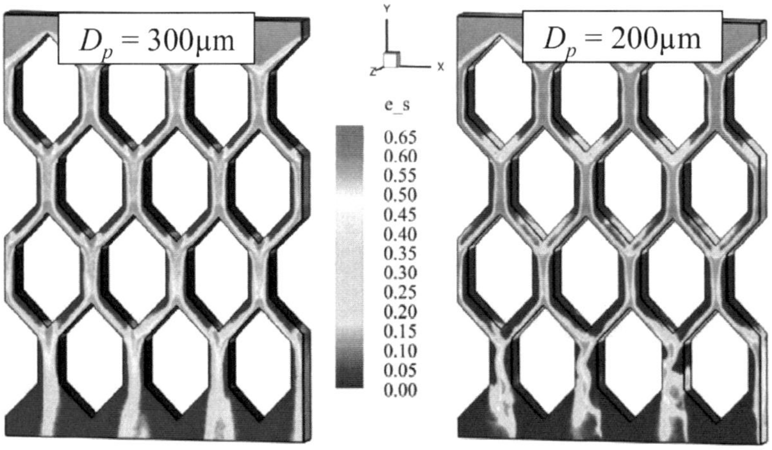

~12 million particles

Comparisons: Local heat transfer coefficient

Particle Diameter

- Good agreement for hexagons in 1st and 2nd rows

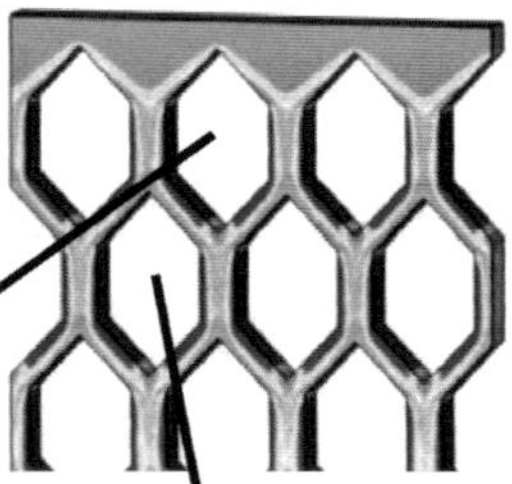

Lines are continuum model
Symbols are DEM

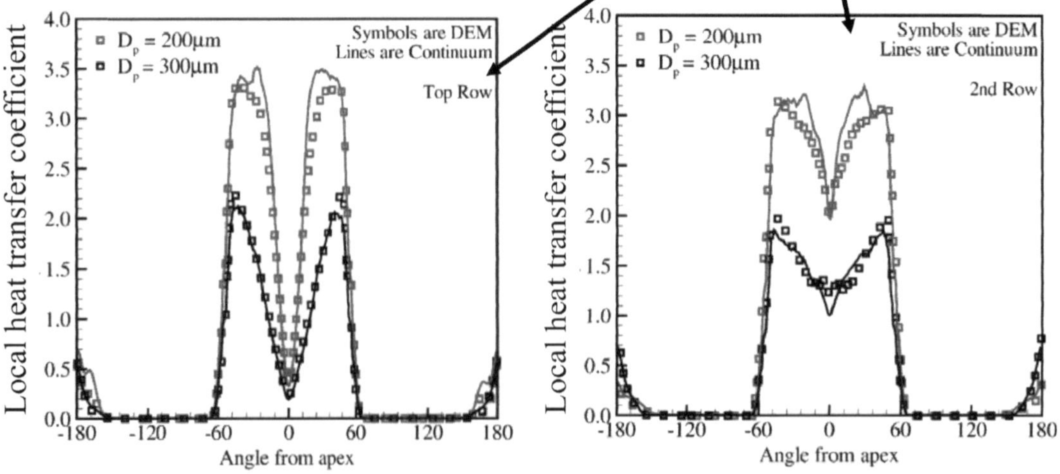

New continuum model vs DEM

Heat transfer most sensitive to

1) Particle size
2) Solids flow rate
3) Apex angle

New continuum model closely matches DEM

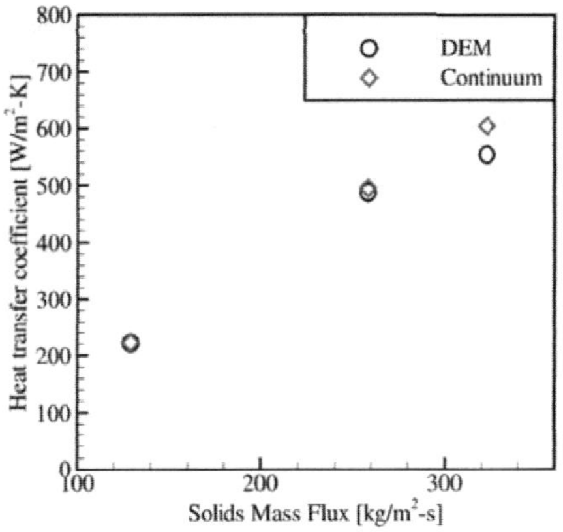

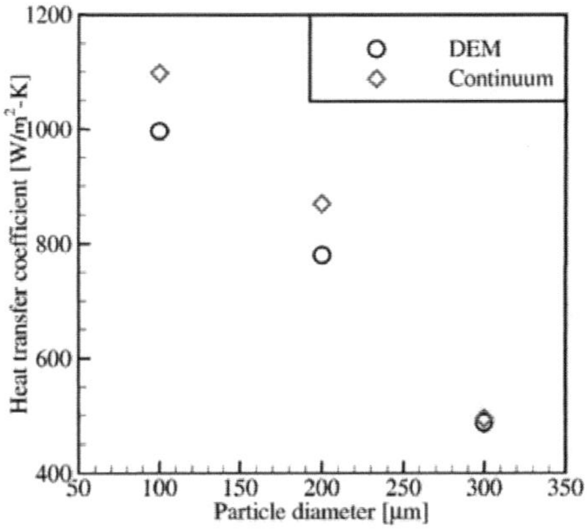

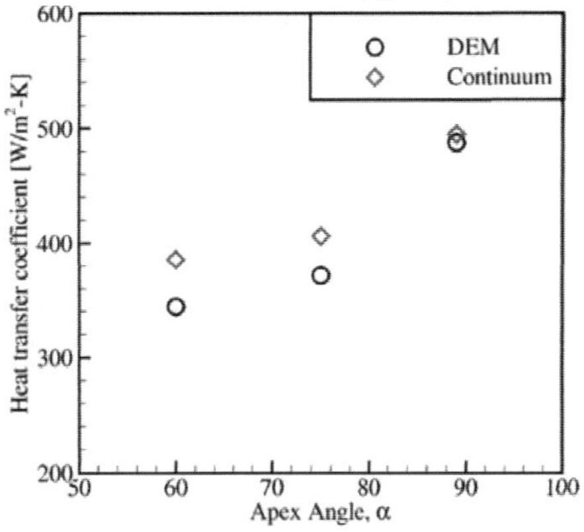

Conclusions

Conductive heat transfer strongly depends on solids concentration and particle-wall contacts

- Previous models were derived for packed beds and are inaccurate for flowing particles

Heat transfer to flowing particles can be counter-intuitive

- Higher velocity -> less heat transfer
- Increased agitation -> less heat transfer (sometimes)

New model accounts for solids concentration via use of 'particle-wall' distribution function

- Accurately matches average heat transfer coefficient (average error less than 5% over all test cases)
- Accurately predicts local heat transfer coefficient

Heat transfer is most sensitive to particle size

- More heat transfer for smaller particles (caveat: no jamming)

22/23

Acknowledgements

Ben Grote (Undergraduate researcher)
University of Colorado

Aaron Lattanzi (PhD candidate)
University of Colorado

SunShot initiative through the U.S. DOE BRIDGE program (grant no. DE-EE0005954).

Oak Ridge Leadership Computing Facility located at Oak Ridge National Laboratory, supported by contract DE-AC05-00OR22725.

The Janus supercomputer, NSF award number CNS-0821794

Backup slides

DEM Simulations of hexagonal tube array

Role of Mass Flow Rate

Mass Flux	Average h [W/m²-K]
258.2 kg/m²-s	487.41
129.1 kg/m²-s	313.28

$$\bar{h} = \frac{\dot{Q}_{tot}}{A_{hex,tot}\left(T_{wall} - T_{s,inlet}\right)}$$

- Less heat transfer at low mass flux (lower concentration)

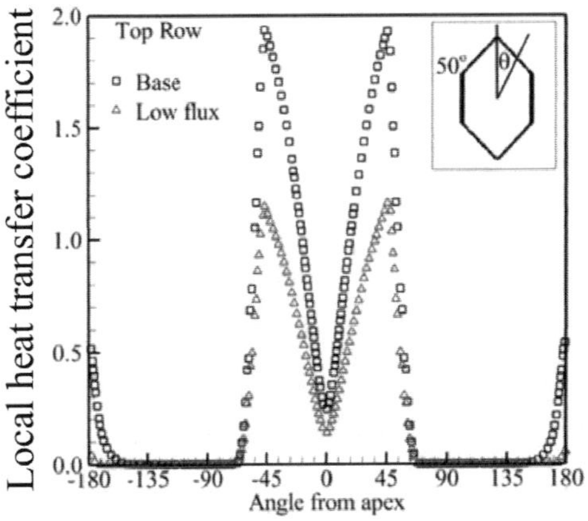

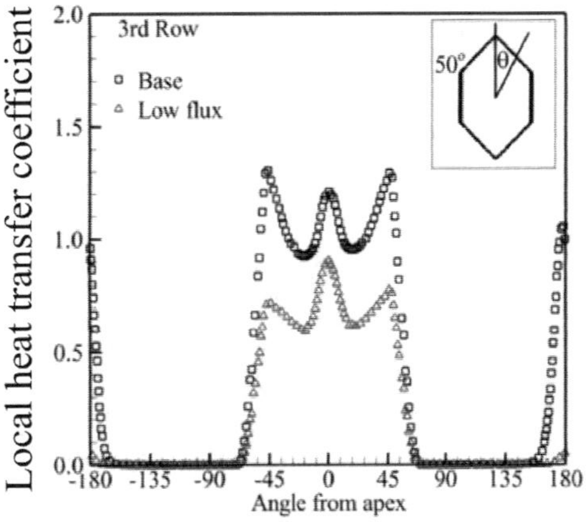

DEM Simulations of hexagonal tube array

Role of Horizontal Gap Spacing

Δx = {5.04cm, 4.80cm, 4.44cm}

Flow chokes for narrow gap

Particle Diameter	300 μm
e (rest. coeff)	0.9
Solids Mass Flux	258.2 kg/m²-s
Inlet Temperature	300 K
Wall Temperature	600 K

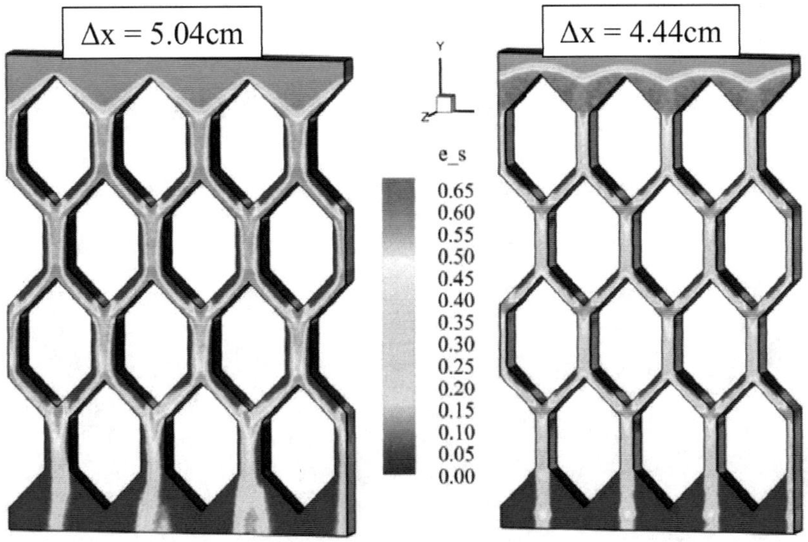

DEM Simulations of hexagonal tube array

Role of Horizontal Gap Spacing

x-spacing	Average h [W/m²-K]
5.04 cm	487.41
4.80 cm	484.90
4.44 cm	584.16

$$\bar{h} = \frac{\dot{Q}_{tot}}{A_{hex,tot}\left(T_{wall} - T_{s,inlet}\right)}$$

- Heat transfer increases ~20% for narrowest gap
  - Flow chokes and improved side wall contact
  - Higher surface conc. in top row

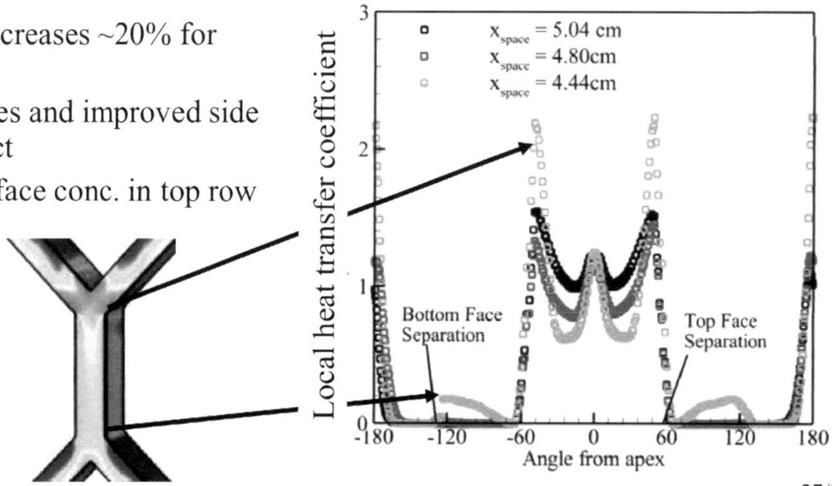

DEM Simulations of hexagonal tube array

Hydrodynamic instabilities in 2-D vs 3-D

- Instabilities develop later in 3-D than 2-D. Still similar to experiments

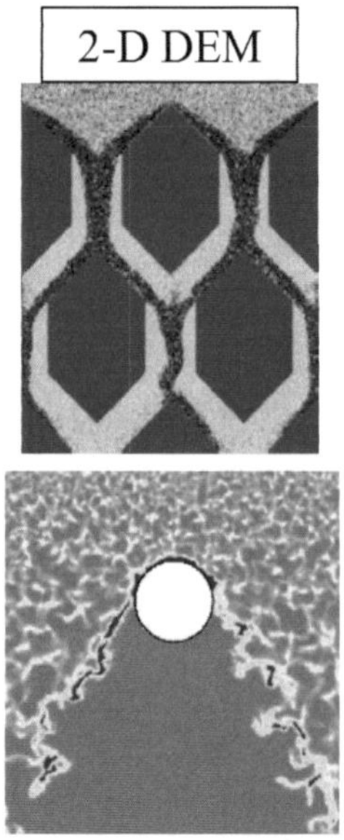

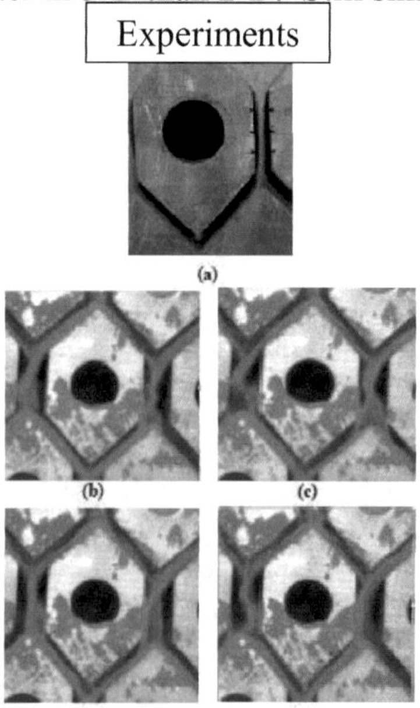

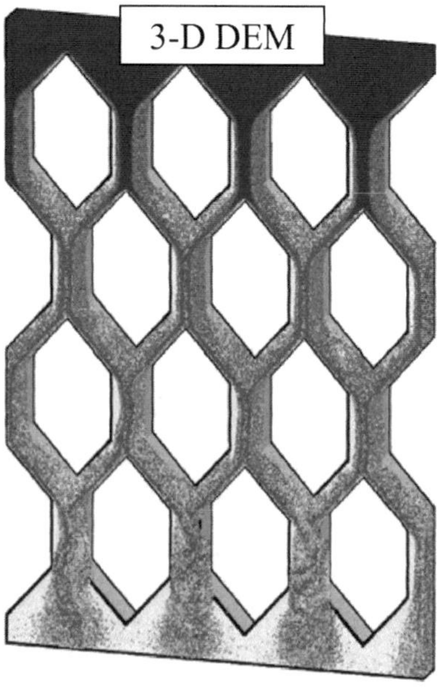

(128c) Topology Optimisation for Optimal Shape Selection of Solar Parabolic Trough Receiver

Ocheme, S., University of Leeds

Originally presented on: 11/9/2015 13:20:00 - 13:45:00

Concentrated Solar thermal technology holds a strong potential for the generation of process heat in the chemical process and allied industries including residential and commercial application. For more than a century the shape of the receiver of a solar thermal parabolic trough had been the simple and easily fabricated circular shape and a non-optimal shape for maximum heat transfer by conduction, convection and radiation heat transfer. In this study, a topology based optimisation methodology was proposed for optimal shape selection from common shapes such as parabola, ellipse, rhombus, equilateral triangle and circular, using collection efficiency and collector element shape as the objective space and ease of manufacturability and cost as the decision space variables or constraints. The result of the research will help power generation and process engineers make decision on the best shape of receiver for process heat production for a minimal cost and ease of manufacture.

Design of solar thermophotovoltaic power generators using a genetic algorithm

John DeSutter, Michael P. Bernardi and Mathieu Francoeur

Radiative Energy Transfer Lab, Department of Mechanical Engineering, University of Utah, Salt Lake City, Utah 84112

I. Introduction

Solar thermophotovoltaic (STPV) power generators are conceptually similar to photovoltaic (PV) solar cells in which solar radiation is converted into electricity. However, STPVs have an intermediate layer between the sun and the STPV cell. This intermediate layer absorbs solar radiation and reradiates this energy towards the STPV cell. The intermediate layer, also called the radiator, allows for the radiation to be spectrally emitted and/or filtered in a way that better matches the cell's absorption characteristics. It has been shown that the theoretical maximum efficiency for a STPV power generator that radiates monochromatically at a frequency matching the absorption bandgap of the cell is 85.4 % [1]. A record STPV efficiency of 3.2% has recently been experimentally demonstrated [2]. Despite this important milestone, solar PV cells still largely outperform STPV power generators.

Many papers have been devoted to the design of selectively emitting radiators for potential use in TPV power generators (e.g., see Ref. [3]). However, these radiators are usually designed without considering radiative, electrical and thermal losses in the cell. Radiation absorbed by the cell with frequency smaller than the bandgap that does not contribute to photocurrent generation is a radiative loss. Electrical losses are caused by electron-hole pairs (EHPs) recombining before reaching the depletion region. Thermal losses are due to absorption of radiation with frequency smaller than the cell's bandgap by the free carriers and the lattice, non-radiative recombination of EHPs and thermalization of radiation with frequency larger than the bandgap. Thermal losses negatively affect STPV performance by increasing the dark current, due to an increase of the cell temperature, opposing the generated photocurrent [4].

In this work, optimum radiator emission spectra maximizing STPV power output and conversion efficiency are designed by accounting for radiative, electrical and thermal losses. This is achieved by coupling a multi-physics model combining thermal radiation, electrical and thermal transport [5] with the publicly available PIKAIA genetic algorithm [6]. The STPV power generator analyzed in this work consists of a cell made of gallium antimonide (GaSb) having an absorption bandgap of 0.723 eV at 293 K, a radiator that has a temperature of 2000 K, and a cell thermal management system characterized by a heat transfer coefficient of 600 $Wm^{-2}K^{-1}$ and a fluid temperature of 293 K. It is shown that the spectra maximizing conversion efficiency and power output greatly differ from each another.

II. Theoretical Framework

When radiative, electrical, and thermal losses are taken into account, determining an emission spectrum maximizing STPV performance is a difficult task, since these three loss mechanisms are coupled with each other [4]. This makes an analytical solution for the optimum spectrum very difficult, if not impossible. To find this optimum spectrum, a multi-physics model combining thermal radiation, electrical and thermal transport in TPV devices, described in Ref. [5], is coupled with the genetic algorithm PIKAIA [6].

The genetic algorithm PIKAIA is an optimization tool utilizing the concept of evolution. PIKAIA determines a set of input parameters maximizing a user defined objective function. In this work, the thermal spectrum emitted by the radiator is discretized into N bands, where the emissivity is assumed to be uniform in a given band. The input parameters in PIKAIA are therefore the N values of emissivity. A given set of N emissivities is a single individual and each emissivity corresponds to a trait of the individual. The total number of sets is the population size. Two objective functions are investigated, namely STPV power output and conversion efficiency. The objective function for each set of emissivities is computed via the TPV model, and PIKAIA ranks each set based on objective function. Parent sets are selected based on rank to create offspring sets that consist of traits from both parent sets. If an offspring set produces a higher objective function than a set in the parent generation, the set in the parent generation is replaced by the offspring set. The population is evolved in this manner over a user specified number of generations in order to determine a set of spectral emissivities that maximize the objective function.

III. Results

The spectrum obtained from PIKAIA maximizing STPV power output for $N = 104$ spectral bands is shown in Fig. 1. The power output, cell temperature and conversion efficiency for this case are 41673 Wm^{-2}, 393 K and 24.4%, respectively. The spectrum takes the form of a step function in which all radiation below 0.685 eV and above 1.082 eV has an emissivity of zero, while all radiation between these limits has the maximum emissivity of 1. Note that at a temperature of 393 K, the bandgap of the GaSb cell reduces to 0.685 eV, thus explaining why the low-energy cutoff is at 0.685 eV instead of 0.723 eV.

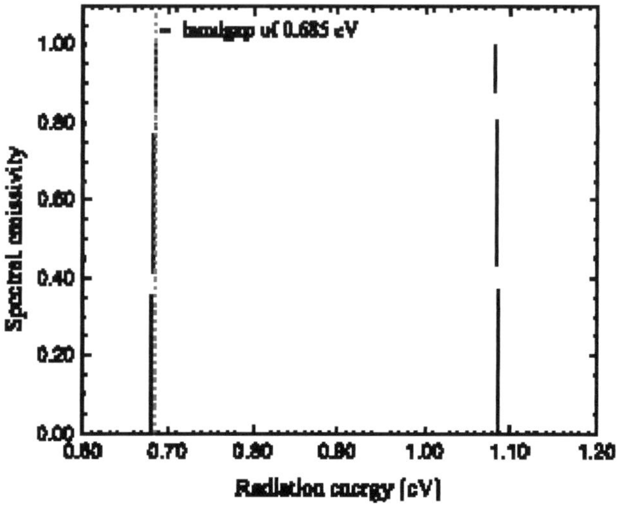

Figure 1. Thermal emission spectrum maximizing STPV power output for a radiator temperature of 2000 K, and a thermal management system having a fluid temperature and a heat transfer coefficient of 293 K and 600 $Wm^{-2}K^{-1}$, respectively.

The high-energy cutoff of 1.082 eV for the optimal spectrum can be explained by analyzing Fig. 2, where the power output, the cell temperature, the current density at maximum power, the bias voltage at maximum power, and the dark current density are plotted as a function of the high-energy cutoff. Note that each curve has been normalized by its own maximum. As the high-energy cutoff increases, the current density increases due to an increasing amount of radiation absorption with energy larger than the cell's bandgap. The increased amount of radiation absorption, however, causes the temperature of the cell to increase due to non-radiative recombination of EHPs and thermalization. This results in an increase of the dark current density which, in turn, causes the bias voltage at maximum power to decrease. Since power output is the product of the current density and the bias voltage, there is a spectral limit beyond which the decrease in bias voltage outweighs the increase in current density. Beyond that limit, the power output starts to decrease with increasing high-energy cutoff.

The spectrum obtained by PIKAIA when maximizing STPV conversion efficiency results in power output, cell temperature and conversion efficiency of 9591 Wm^{-2}, 302 K and 38.9%, respectively. Interestingly, this spectrum is not simply monochromatic at the cell's bandgap. Instead, the spectrum takes the form of a narrow step function, where radiation below 0.723 eV and above 0.761 eV has an emissivity of zero, while radiation between these limits has the maximum emissivity of 1.

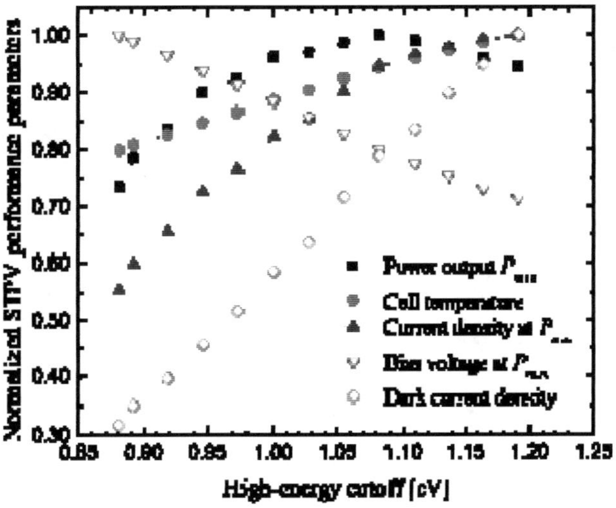

Figure 2. Normalized power output, cell temperature, current density at maximum power, bias voltage at maximum power, and dark current density as a function of the high-energy cutoff of the radiator spectrum.

Figure 3 shows STPV power output obtained from different radiators: blackbody, tungsten, optimum spectrum A obtained by maximizing power output and optimum spectrum B obtained by maximizing conversion efficiency. It can be seen that the power output produced by the optimum spectrum A is nearly twice that of the blackbody and tungsten radiators, and exceeds the optimum spectrum B by a factor of four. In addition, the cell temperature of 505 K with a blackbody radiator greatly exceeds that of the tungsten, optimum spectrum A and optimum spectrum B cell temperatures of 380 K, 393 K, and 302 K, respectively. The high thermal losses for the blackbody radiator result in a very low conversion efficiency of 8.44% compared to that of tungsten, optimum spectrum A and optimum spectrum B values of 19.3%, 24.4% and 38.9%, respectively.

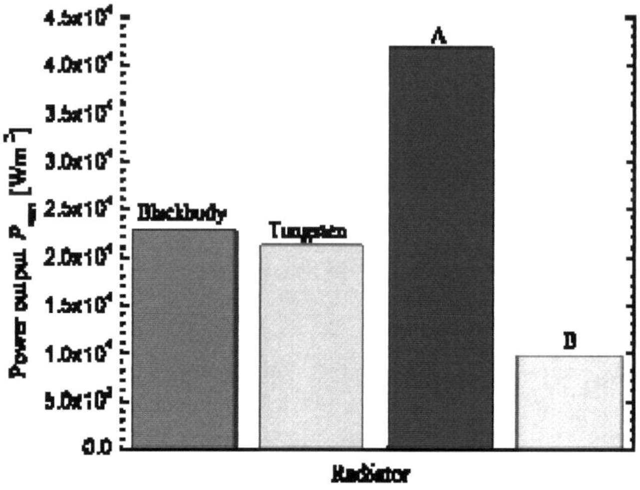

Figure 3. Comparison of power output obtained with various STPV radiators: blackbody, tungsten, optimum spectrum A and optimum spectrum B.

IV. Conclusions

This work shows the importance of taking radiative, electrical and thermal losses into account when designing optimal STPV power generators. The next step is to examine how other system level variables such as the temperature of the fluid and the heat transfer coefficient in the thermal management system affect the radiator emission spectrum when maximizing power output and conversion efficiency.

V. References

[1] N.-P. Harder, P. Wurfel, *Semicond. Sci. Technol.* **18**, S151, 2003.

[2] A. Lenert, D. M. Bierman, Y. Nam, W. R. Chan, I. Celanovic, M. Soljacic, E. N. Wang, *Nat. Nanotechnol.* **9**, 126, 2014.

[3] B. Zhao, L. Wang, Y. Shuai, Z. M. Zhang, *Int. J. Heat Mass Tran.* **67**, 637, 2013.

[4] M.P. Bernardi, O. DuprŽ, E. Blandre, P.-O. Chapuis, R. Vaillon, M. Francoeur, *Sci. Rep.*, under review, 2015.

[5] M. Francoeur, R. Vaillon, M.P. MengŸ□, *IEEE T. Energy Conver.* **26**, 686-698, 2011.

[6] PIKAIA: http://www.hao.ucar.edu/modeling/pikaia/pikaia.php. May 10 2015.

432942 Near-Field Enhanced Thermionic Energy Conversion for Concentrated Photovoltaic Power Generation

Monday, November 9, 2015: 2:10 PM
259 (Salt Palace Convention Center)

Mohammad Ghashami, Mechanical Engineering Department, University of Utah, Salt Lake City, UT and Keunhan Park, Mechanical Engineering, the university of UTAH, Salt Lake City, UT

This abstract presents the direct conversion of excessive heat from a concentrated photovoltaic power generator into the electrical power by developing near-field enhanced thermionic energy conversion (NETEC). The excess heating of the concentrated photovoltaics (CPV) is the main source of waste energy that should be effectively removed to maintain the performance of photovoltaic cells. To make use of waste heat generated in the (CPV) system, we propose two innovative approaches integrated with thermionic emission: (1) combination of photovoltaic and thermionic effects to enhance electron emission from a cathode [1-4], and (2) use of near-field thermal radiation, typically several orders of magnitude greater than blackbody thermal radiation, to boost photovoltaic emission of electrons. This hybrid model takes advantage of the combination of near-field photo-excitation enhancement with thermionic emission which will significantly improve the energy conversion efficiency and output power for the cases where cathode temperature is much lower than conventional thermionic energy converters. A thermal emitter and a low-bandgap semiconductor cathode are separated by a sub-wavelength gap distance to allow the transfer of a significant amount of thermal radiation between the two due to near-field effects. Near-field thermal radiation enhances the concentration of electrons in the conduction band of the cathode due to the combination of photoexcitation and thermal excitation, leading to the enhancement of electrical current from the cathode to the anode. This hybrid system will be remarkably more efficient with higher power throughput compared to other current direct energy conversion systems, such as thermoelectric, thermophotovoltaic, and thermionic systems [5-7].

Figure 1(a) shows the schematic configuration of the NETEC system, where a nanoscale gap is maintained between cathode and emitter to allow strong enhancement in thermal radiative transfer. The thermal energy required to maintain the emitter at a high temperature can be supplied by the excessive heat coming from solar concentrated energy. The energy diagram in Fig.1(b) illustrates the work functions and energy barriers of this system, where the output voltage is basically given by the difference between the cathode and anode work functions plus any external voltage, *Vex* V_{ex}.

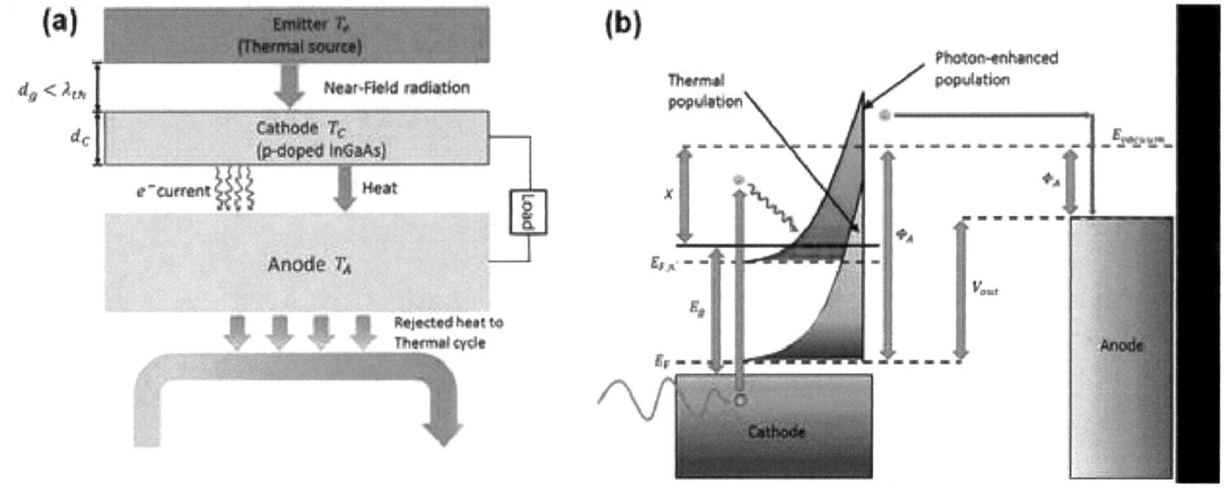

Figure SEQ Figure * ARABIC 1. (a) Schematic of a NETEC structure showing the heat transfer and carrier transport mechanisms in a plain system. (b) Energy diagram showing the work functions and energy barriers.

In order to calculate the current density of the cathode, by following a similar procedure to calculation of simple thermionic current [1] and by taking into account the effect of photoexcited electrons in the conduction band, following equation will be obtained [1].

*JC=4πem*k2h3TC2 exp-EC-EF,n+χkTC=ATC2 exp-□C-EF,n-EFkTC* *(1)*

$$J_C = \left(\frac{4\pi e m^* k^2}{h^3}\right) T_C^2 \exp\left[\frac{-(E_C - E_{F,n} + \chi)}{kT_C}\right] = A T_C^2 \exp\left[\frac{-\left(\phi_C - (E_{F,n} - E_F)\right)}{kT_C}\right] \quad (1)$$

Hence, the output power of a NETEC system can be calculated analogous to thermionic converters, by multiplying the net current density, *JC-Ja* $J_C - J_a$, and the output voltage,

PNETEC=JnetVnet=JC-Ja□C-□a+Vex
(2)

$$P_{NETEC} = J_{net} V_{net} = (J_C - J_a)(\phi_C - \phi_a + V_{ex}) \quad (2)$$

Further, the efficiency of the NETEC system, *ηsys* η_{sys} can be defined as the ratio of the output electrical power, *Pel* P_{el} over the total input energy of the system, which in this case, is the summation of the total power emitted by the thermal emitter (calculated from near-field radiative transfer equations), *Pemitter* $P_{emitter}$ and the net thermal energy needed to maintain the cathode temperature, *QCathode* $Q_{Cathode}$, *ηsys=PelPemitter+QCathode*

(3)

$$\eta_{sys} = \frac{P_{el}}{P_{emitter} + Q_{Cathode}} \qquad (3)$$

The operating temperature of the cathode is a decisive parameter in designing an efficient NETEC system, since the thermal energy of the cathode which is directly related to its temperature, determines the average velocity of the carriers within the material. Thus, by decreasing the electron affinity of the material, χ X, cathode should be able to operate in lower temperatures. The effects of electron affinity on the cathode output current, the electrical power and the system efficiency are shown in Fig.2(a) , (b) and (c). As it can be seen in Fig.2(c), for any electron affinity, there exist an optimum temperature which results in the maximum system efficiency, where by increasing the electron affinity, intuitively this optimum temperature will raise as accordingly. So, by wisely manipulating the electron affinity of the cathode through some surface operations to manage it due to its operating temperature, we can obtain a remarkable output power.

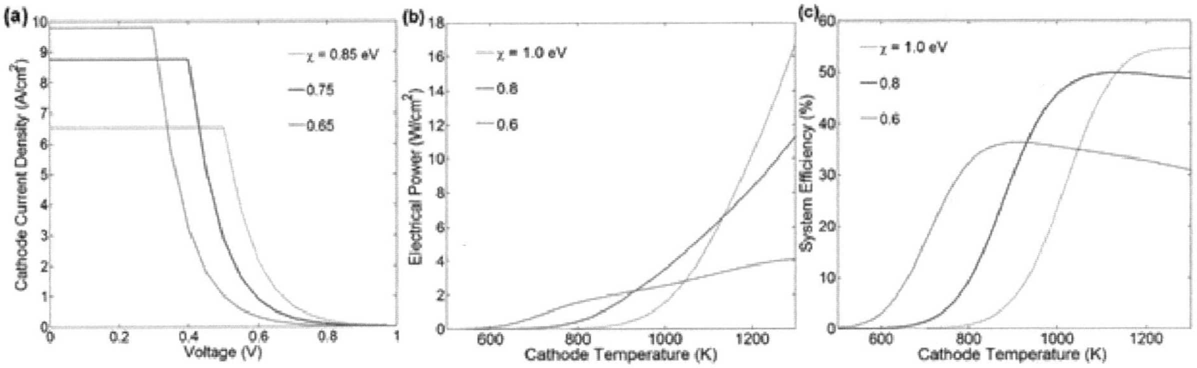

Figure SEQ Figure * ARABIC 2. (a) J-V characteristics of a NETEC system for different electron affinities. (b) Electrical power as a function of cathode temperature. (c) System efficiency for different electron affinities and cathode temperature.

To summarize, in this study a novel hybrid energy conversion system is introduced by combination of the near-field thermal radiation and conventional thermionic energy conversion systems. Near-field enhanced thermionic energy conversion (NETEC) system benefits from photoexcitation due to incident near-field radiation and also thermal energy of the semiconductor cathode in order to directly generate electricity. We have illustrated the remarkable enhancement in output power and system efficiency achievable using this new concept for the first time. Most importantly, many challenges faced in direct energy conversion schemes such as near-field thermophotovoltaic systems or thermionic converters, being an essentially high temperature or low power throughput, can be addressed and resolved using this method.

[1] Schwede, Jared W., et al. "Photon-enhanced thermionic emission for solar concentrator

systems." *Nature materials* 9.9 (2010): 762-767.

[2] Segev, Gideon, Yossi Rosenwaks, and Abraham Kribus. "Loss mechanisms and back surface field effect in photon enhanced thermionic emission converters." *Journal of Applied Physics* 114.4 (2013): 044505.

[3] Wang, Yuan, et al. "Parametric design criteria of an irreversible vacuum thermionic generator." *Journal of Applied Physics* 114.5 (2013): 053502.

[4] Zhuravlev, A. G., A. S. Romanov, and V. L. Alperovich. "Photon-enhanced thermionic emission from p-GaAs with nonequilibrium Cs overlayers." *Applied Physics Letters* 105.25 (2014): 251602.

[5] Krier, Anthony. *Mid-infrared semiconductor optoelectronics*. Springer, 2006.

[6] Chen, Gang. "Nanoscale heat transfer and nanostructured thermoelectrics."*Components and Packaging Technologies, IEEE Transactions on* 29.2 (2006): 238-246.

[7] Lee, Jeong Ik, et al. "Size effect of nanometer vacuum gap thermionic power conversion device with CsI coated graphite electrodes." *Applied Physics Letters* 95.22 (2009): 223107.

414310 A Parallel, Open Source, Grey Body Monte Carlo Ray Tracer with Graphical User Interface for Solar Simulator Characterization

Abstract:

Solar-thermal technologies harness energy with heliostats, heliostats that heat a boiler or reactor with reflected sunlight. This approach has been successfully used to drive biomass gasification, metals reduction and electrical generation in a renewable manner [1-3]. Solar-thermal research and development is accelerated by solar simulators: assemblies of high power lamps whose irradiance mimics concentrated sunlight [4]. These platforms provide controlled environments for the evaluation new solar-thermal technologies at the laboratory scale. To date seven solar simulators are available worldwide for academic studies [4]. Solar simulators can be analyzed using Monte Carlo ray tracing, a probabilistic approach to modeling radiative heat exchange [5]. Within this regime the fates of radiation are determined by following the paths of individual light rays *in silico*. Rays from a given lamp are traced through diffuse and specular reflections to their ultimate absorption. Several Monte Carlo computer programs have been written for solar simulator characterization. However, most were tailored to the analysis of specific problems [4, 6]. A general Monte Carlo ray tracing code for academic use, VEGAS, features attributes common in scientific programming [7]. Specifically, interaction with VEGAS is purely through text inputs and post-processing is left to the user. The program was written in Fortran90, a compiled computer language wherein modern software practices are emulated [8]. Interpreted programming languages that support object oriented coding have gained traction in scientific computation [9]. Program prototyping is often faster in these languages relative to compiled languages, although at runtime interpreted code may execute more slowly [10]. The use of parallel computing has the potential to ameliorate the latter deficiency. We describe a new, parallel, Monte Carlo Ray ray tracer written in the interpreted Matlab programming language. ParallelMatlabTrace features a graphical user interface and integrated post-processing tools for solar simulator analysis. This code was used to analyze reradiation from two cylindrical calorimeters for solar simulator calibration. Monte Carlo results were coupled with a finite volume simulation to describe worst-case thermal reradiative losses. Natural convection was not explored given its negligible effects in prior studies.

1. Wieckert, C., et al., *A 300kW solar chemical pilot plant for the carbothermic production of Zinc.* Journal of solar energy engineering, 2007. **129**(2): p. 190-196.

2. Lichty, P., et al., *Rapid high temperature solar thermal biomass gasification in a prototype cavity reactor.* Journal of solar energy engineering, 2010. **132**(1): p. 011012.

3. Behar, O., A. Khellaf, and K. Mohammedi, *A review of studies on central receiver solar thermal power plants.* Renewable and Sustainable Energy Reviews, 2013. **23**: p. 12-39.

4. Bader, R., S. Haussener, and W. Lipiñski, *Optical Design of Multisource High-Flux Solar Simulators.* Journal of solar energy engineering, 2015. **137**(2): p. 021012.

5. Howell, J., *The Monte Carlo method in radiative heat transfer.* Journal of Heat Transfer, 1998. **120**(3): p. 547-560.

6. Martinek, J., C. Bingham, and A.W. Weimer, *Computational modeling and on-sun model validation for a multiple tube solar reactor with specularly reflective cavity walls. Part 1: Heat transfer model.* Chemical engineering science, 2012. **81**: p. 298-310.

7. Petrasch, J.r. *A free and open source Monte Carlo ray tracing program for concentrating solar energy research.* in *ASME 2010 4th International Conference on Energy Sustainability*. 2010. American Society of Mechanical Engineers.

8. Decyk, V.K., C.D. Norton, and B.K. Szymanski, *How to Express C++ Concepts in Fortran90.* Scientific Programming, 1997. **6**(4): p. 363-390.

9. Prabhu, P., et al. *A survey of the practice of computational science*. in *State of the Practice Reports*. 2011. ACM.

10. Prechelt, L., *An empirical comparison of seven programming languages.* Computer, 2000. **33**(10): p. 23-29.

Study of Plenum to Plenum (P2P) Natural Circulation Phenomena in a Dual Channel Scaled Module of Very High Temperature Reactor Design By Using CFD

M. T. Kao[1], P. Jain[2], S. Usman[3], I. A. Said[4], M. M. Taha[4], M. H. Al-Dahhan[4], and Rizwan-Uddin[1]

1) Department of Nuclear, Plasma and Radiological Engineering, University of Illinois at Urbana-Champaign.
2) Oak Ridge National Lab.
3) Department of Mining and Nuclear Engineering, Missouri S&T, Rolla.
4) Department of Chemical Engineering, Missouri S&T, Rolla.

Presenting Author: Mahmoud Taha Moharam (M. M. Taha)

MISSOURI S&T

MISSOURI UNIVERSITY OF SCIENCE AND TECHNOLOGY

Presentation Outline

1. Introduction
2. Motivation
3. Objectives
4. Work plan
5. Preliminary Computational Simulations
6. Missouri S&T Scaled-down Facility (MS&T-F)
7. CFD Simulations
8. Remarks
9. Acknowledgement

2

Introduction

- The International forum in 2000 have released the Very High Temperature Reactors (VHTRs) is one of the most leading candidate for Next Generation Nuclear Plants (NGNPs).
- The design has high fuel efficiency for electricity production and because of the high gas coolant outlet temperature is optimal for industrial applications or commercial hydrogen production.

3

MISSOURI UNIVERSITY OF SCIENCE AND TECHNOLOGY

Introduction "cont."

- There are two versions of VHTRs:
  1. Pebble Bed Reactor (PBR)
  2. Prismatic Modular Reactor (PMR)
- The Gas-Turbine Modular Helium-cooled Reactor (GT-MHR), also termed Modular High Temperature Gas Reactor (MHTGR) is a candidate for the DOE-funded NGNPs prismatic modular reactors.

4

Missouri University of Science and Technology

Introduction "cont."

- MHTGR (prismatic modular) design features:

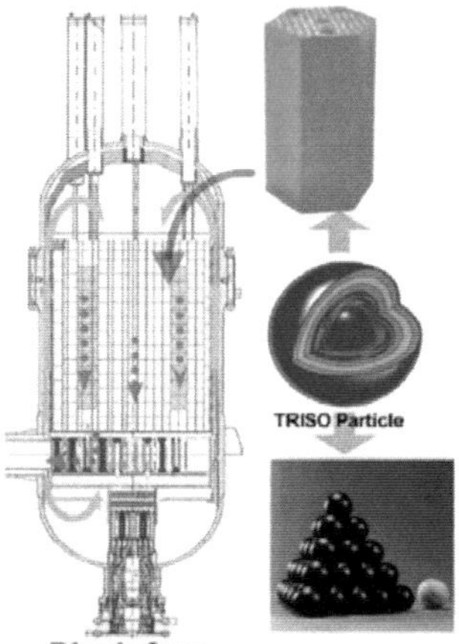

Figure 1. Features of Prismatic VHTR (yoon et al. 2011)

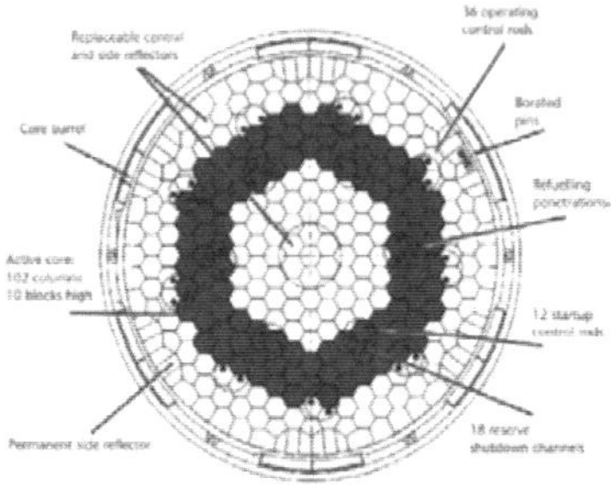

Figure 2. Core of prismatic CHTR (LaBar et al. 2004, Southworth et al. 2004)

5

Missouri University of Science and Technology

Motivation

- Computational Fluid Dynamics (CFD) has received a great deal of attention for understanding and predicting the flow behavior in VHTR, **BUT sufficient validation of these codes has not been carried out.** (Lee et al. 2014, McVay et al. 2015)
- Once CFD codes have been validated, computational tools enable process visualization and understanding and hence design optimization.

6

Objectives

- *Long term goal*: Obtaining benchmark experimental data to validate the CFD codes for the design and analyses of PMR-VHTR.

"benchmark data is needed to validate CFD simulations while there were no experimental test facilities to provide this"

McVay et al. 2015

Objectives

- *At this stage*: Using current preliminary CFD simulations to identify the most suitable locations to place the sensors of the novel sophisticated measurement techniques employed on Missouri S&T scaled-down Facility (MSTF).
- Techniques to be used:
  1. Heat flux foil-sensors
  2. Hot wire anemometry
  3. Gas tracer

Work plan

Preliminary CFD simulations

↓

Experimental setup construction

↓

New simulation

MISSOURI UNIVERSITY OF SCIENCE AND TECHNOLOGY

Preliminary CFD Simulations

- CFD studies carried out to determine the best locations for instrumentation by using two different codes;
  1. STAR-CCM+
  2. COMSOL (confirmatory analysis on primary computational results obtained by STAR-CCM+)
- Flow conditions are simulated for coolant Helium for operational conditions at 0.8 MPa and 4.2 MPa.

Missouri S&T Facility (MS&T-F)

- The same scaling criteria used for building the High Temperature Test Facility (HTTF) at Oregon State University (OSU) with reference to MHTGR are employed.
- These scaling criteria have shown a successful similarity between HTTF and MHTGR. (Brian M. King 2012, R.R. Schultz et al. 2010)

MSTF "cont."

- MS&T dual-channel Facility *"Phase-I"* is a 1:4 scaled down facility based on OSU-HTTF design as shown below.

Table 1. Dimensions of MSTF

Parameter	Missouri S&T Facility
Tube Diameter (inch)	0.625 (same as HTTF & MHTGR)
Tube length(inch)	23.228 (1/4 of HTTF channel length)
Core diameter(inch)	11.78 (1/4 of HTTF core diameter)
Upper plenum height (inch)	9.41 ((1/4 of HTTF plenum height)
Outer vessel diameter (inch)	15 (1/4 of HTTF vessel diameter)
Number of channels	Two (one for up-ward and one for down-ward flow)

MSTF "cont."

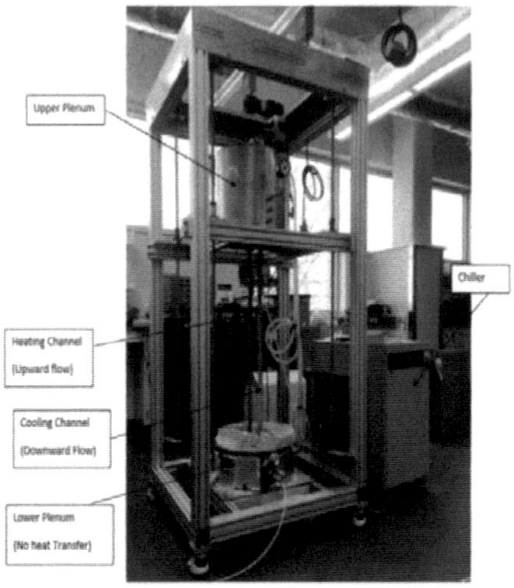

Figure 3. Physical picture of MSTF experimental setup

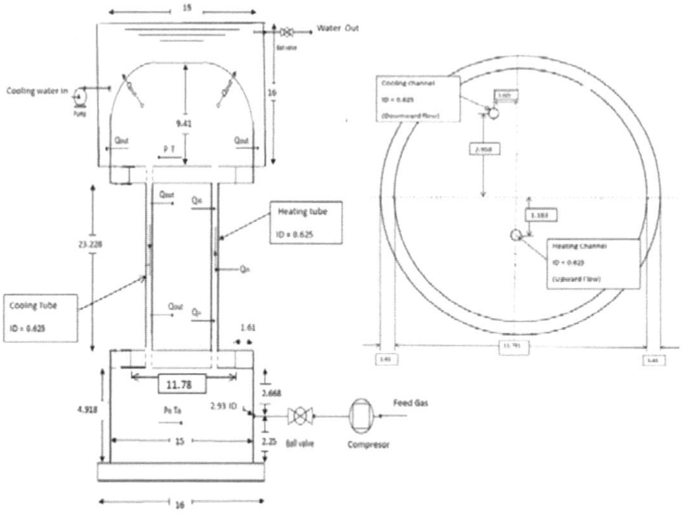

Figure 4. Schematic diagram of MSTF

MISSOURI UNIVERSITY OF SCIENCE AND TECHNOLOGY

CFD simulations

- CFD simulations have been carried out after constructing the experimental setup and both preliminary experimental and computational results have been compared.
- These simulations helps in make modifications on the experimental setup design to decide the final setup design.

Missouri University of Science and Technology

CFD simulations *"cont."*

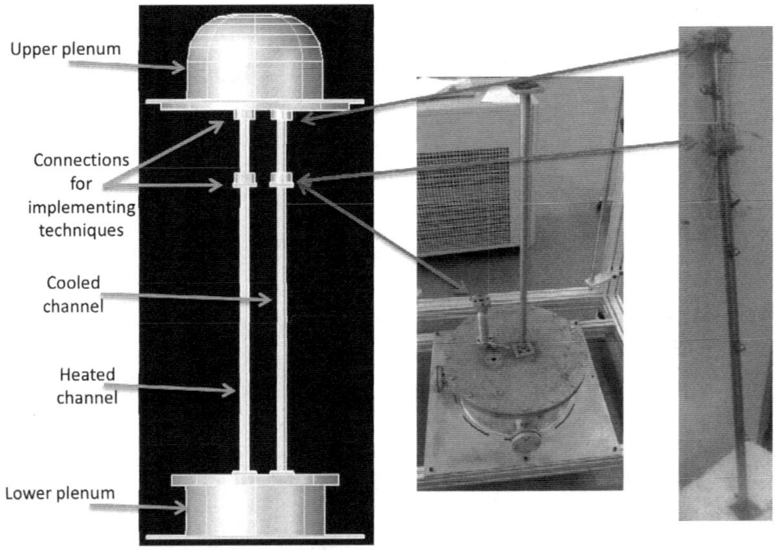

Figure 5. New CAD geometry for experimental setup

15

Missouri University of Science and Technology

CFD simulations *"cont."*

- Initial conditions

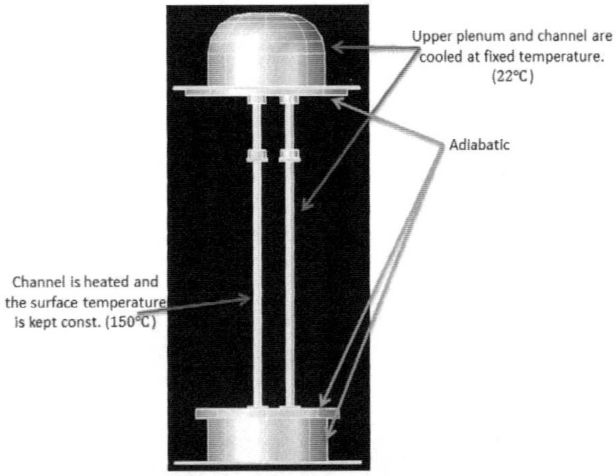

Figure 6. Initial conditions used for new CFD simulations

16

CFD simulations "cont."

- Grid-dependent test (mesh sensitivity analysis)

Table 2. Results obtained at different number of meshing cells

No. of meshing cells	8.6 million	9 million	12 million
Hot channel velocity at mid plane (m/s)	0.775	0.770	0.771
Cold channel velocity at mid plane (m/s)	0.634	0.637	0.637
Heat transfer rate at hot channel (W)	130.29	126.78	126.90
Heat transfer rate at cold channel (W)	8.49	8.29	8.35
Heat transfer rate at top surface (W)	121.89	118.56	118.56

7. CFD simulations "cont."

- Meshing

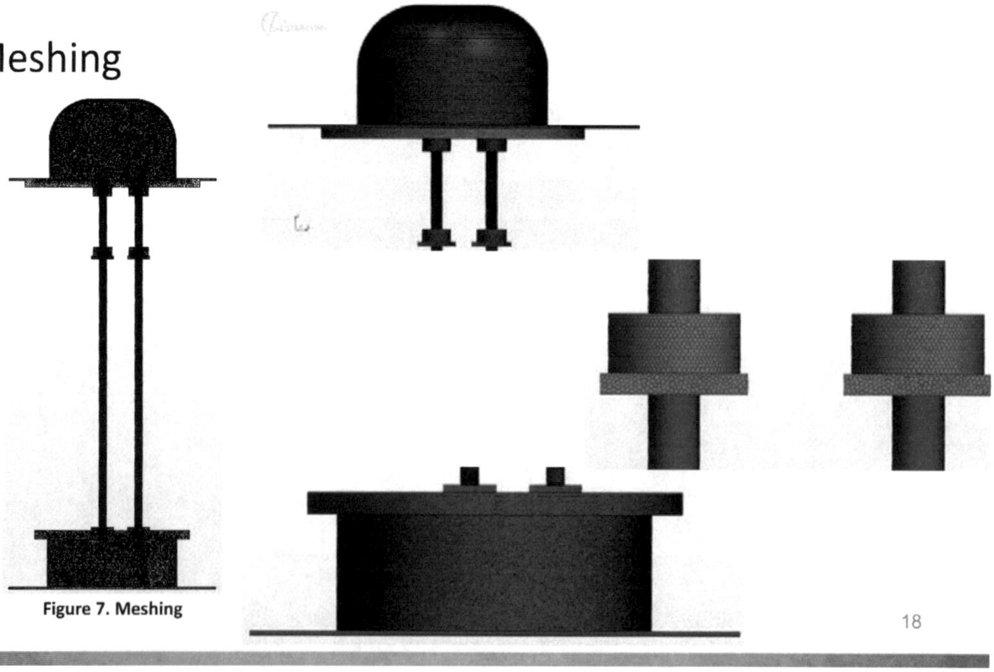

Figure 7. Meshing

CFD simulations *"cont."*

- Simulation setup

Table 3. Simulation setup summary

Fluid	Air, 100 psi
Solid	Stainless steel, alloy 304
Meshing cell numbers	9 million
Turbulence model	Realizable k-epsilon model with two-layer wall treatment
Heated channel surface temperature	150 °C
Cooled channel surface temperature	25 °C
Top surface temperature (upper plenum)	25 °C
Simulation state	3D, steady

19

MISSOURI UNIVERSITY OF SCIENCE AND TECHNOLOGY

CFD simulations *"cont."*

- Temperature distribution profile results:

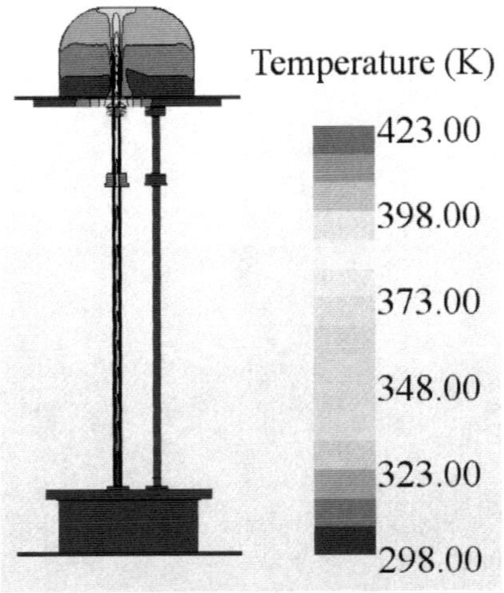

Figure 8. Temperature distribution profile

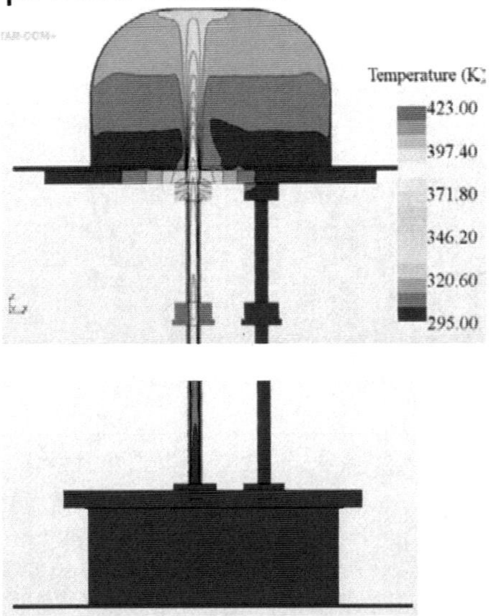

20

CFD simulations "cont."

- Velocity vectors:

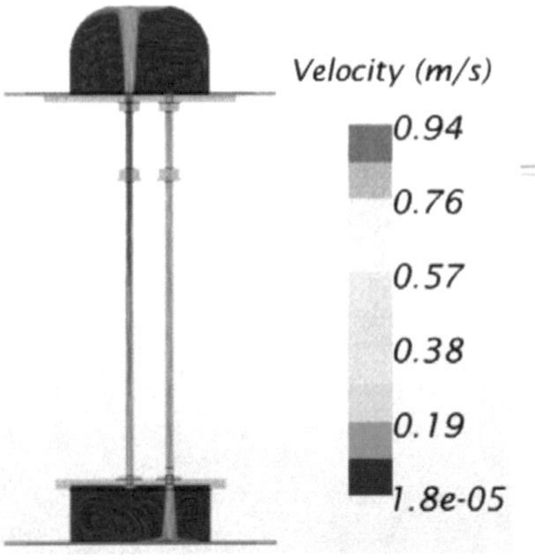

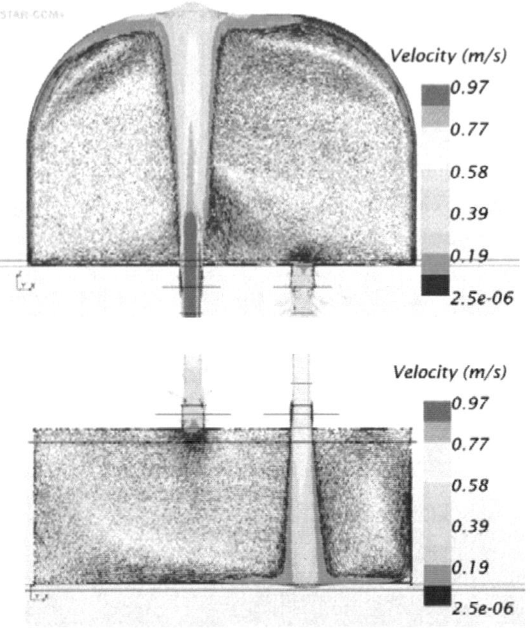

Figure 9. Velocity visualization

21

Missouri University of Science and Technology

CFD simulations "cont."

- Axial temperature profile results:

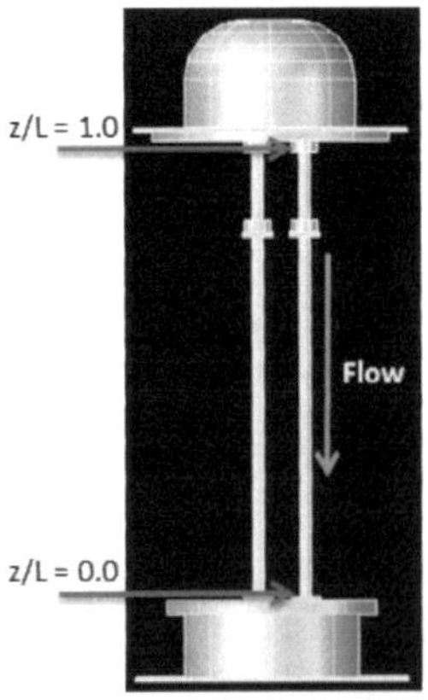

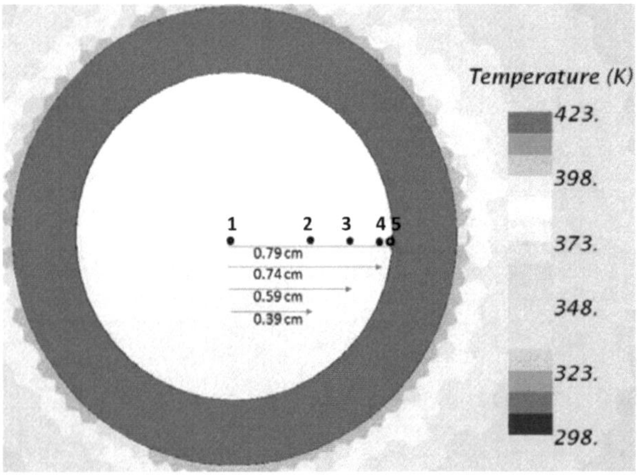

Figure 10. Locations of lines at which temperatures are monitored

22

CFD simulations *"cont."*

- Axial temperature profile results:

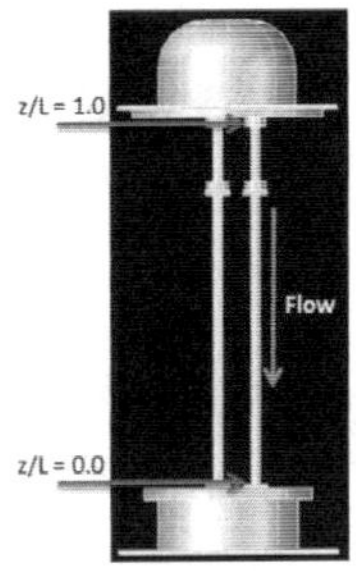

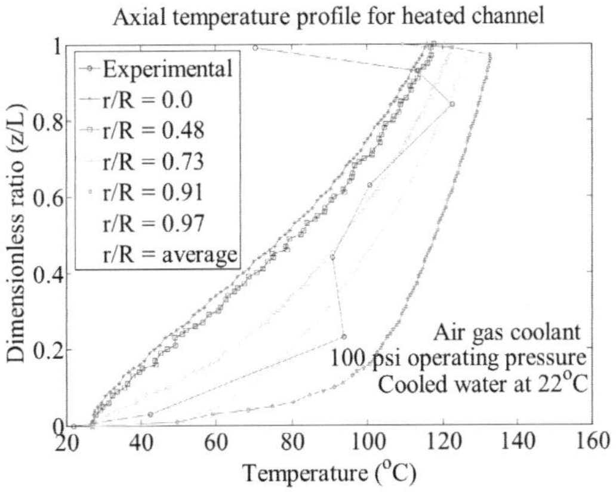

Figure 11. Comparison between axial temperature profiles along different positions for heated channel

Missouri University of Science and Technology

CFD simulations *"cont."*

- Axial temperature profile results:

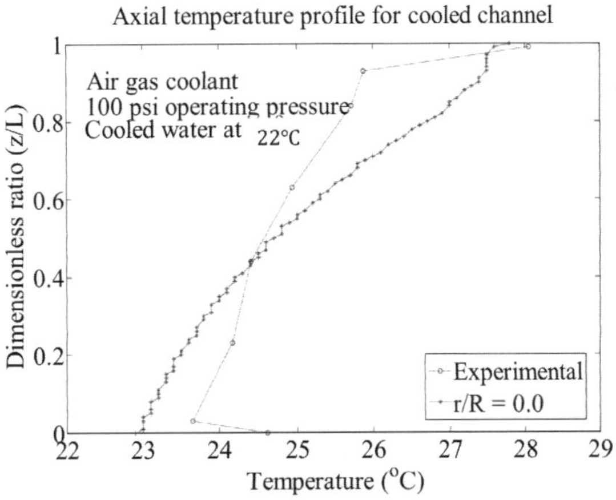

Figure 12. Axial temperature profiles along different positions for cooled channel

Remarks

- Preliminary and new simulations have shown that:
  1. There is a significant effect of radial thermocouples location especially along heated channel on the temperature profile to be obtained.
  2. Hot wire anemometry technique could be placed at different axial and radial positions along the channels.
  3. The gas dynamics at top and bottom parts of channels as well as both plena is needed to be understood by implementing gas tracer technique i.e., back mixing if any.

25

Missouri University of Science and Technology

Remarks "cont."

4. STAR-CCM+ laminar flow model shows convergence only for 0.8 MPa with helium and not with other gases.
5. The flow condition is complicated:
   - It might be laminar (based on the fluid to be used)
   - It is like turbulent jet flow from heated channel to upper plenum

And hence, the experimental data are also needed in guiding selecting the appropriate turbulent model for simulations.

26

Acknowledgment

The authors acknowledge the financial support provided by the U.S. Department of Energy project number (NEUP 13-4953 (DE-NE0000744)) for the 4th generation nuclear energy, which made this work possible.

THANK YOU ...

Team members	Emails
M.T. Kao	mkao2@illinois.edu
P. Jain	jainpk@ornl.gov
S. Usman	usmans@mst.edu
I.A. Said	iasvx4@mst.edu
M.M. Taha	mmt8t6@mst.edu
*M.H. Al-Dahhan (Project **PI**)*	aldahhanm@mst.edu
Rizwan-Uddin	rizwan@illinois.edu

CFD team & Experimental team

421917 Comprehensive Thermodynamic Model for Aqueous Nitric Acid and Sodium Nitrate Solution with Electrolyte NRTL Equation

Abstract:

Comprehensive Thermodynamic Model for Aqueous Nitric Acid and Sodium Nitrate Solution with Electrolyte NRTL Model

Maximilian B. Gorensek [a]

Daniel P. Lambert [a]

Meng Wang [b]

Chau-Chyun Chen [b]

[a] Savannah River National Laboratory, Aiken, SC 29808

[b] Texas Tech University, Lubbock, TX 79409

Abstract

Nitric acid is widely used as a solvent in reprocessing of spent nuclear fuel. It is typically neutralized with sodium hydroxide, leading to significant amounts of sodium nitrate in nuclear wastes. To support heat and mass balance calculations and process simulation for nuclear waste treatment, a comprehensive thermodynamic model is developed for the nitric-acid-sodium nitrate-water ternary system and its subsystems. Based on the symmetric electrolyte NRTL (eNRTL) activity coefficient model, the present work takes into account complete dissociation of sodium nitrate and partial dissociation of nitric acid in aqueous solution. With a maximum of three temperature coefficients for each eNRTL binary interaction parameter, the model provides an accurate and thermodynamically consistent representation for phase equilibrium properties such as vapor pressure, boiling point, dew point and salt solubility; calorimetric properties such as enthalpy and heat capacity; and speciation properties such as molecular nitric acid content. The model is validated with data covering temperatures up to 379.15 K and nitric acid concentrations up to pure acid for the nitric acid-water binary, temperatures up to 473.15 K and sodium nitrate concentrations up to saturation for the sodium nitrate-water binary, and sodium nitrate concentrations up to 0.21 mole fraction and nitric acid concentrations up to 0.3 mole fraction for the nitric acid-sodium nitrate-water ternary system. The model is being extended to cover additional ionic and organic species present in liquid waste at the Savannah River Site.

A Tookit Approach for Simulation of Nuclear Facilities

Valmor F. de Almeida
Oak Ridge National Laboratory, Oak Ridge, TN 37831-6181

Abstract

This presentation will cover the development of a toolkit for modeling and simulation of coupled processes in a nuclear facility; in particular for nuclear fuel reprocessing. Interest in modeling and simulation of facilities arises from different perspectives and often require the capability of experimenting with various flowsheets for different processes. This prompts a toolkit approach for assembling an application from various computational modules which simulate significant processes. The ability to couple these modules in different combinations enables the creation of simulations for different reprocessing facilities. Of particular interest is the time coupling of these individuals modules that synchronizes data exchange in the overall application. The work pursued here is at the plant-level scale. The resulting computational modules are designed with common external and internal behaviors to allow for integration and modular plant design. In addition, the approach supports the development of a reduced-order model wherein subscale phenomena can be taken into account on a more systematic way. The resulting balance equations for the state variables of the network model provide the means for evolving the state of the system and for predicting quantities of practical interest.

An example of aqueous reprocessing will be described with simulations coupling a network of modules involving the head-end and separations areas including: fuel chopping, storage, dissolution, condenser, scrubber, and solvent extraction. Time dependent results for various flow streams will provide an idea of how extensible and scalable the method currently is. Additional simulation power can be obtained by the use of parallel computing which applies well to the approach described.

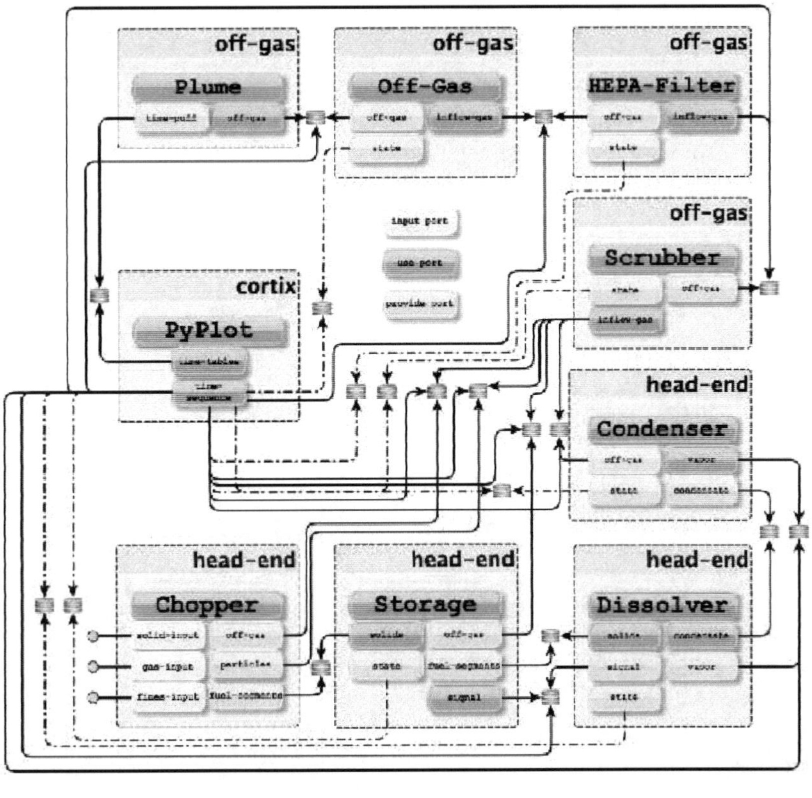

426099 Microsecond-Length Atomistic Simulations of Plasma-Facing Materials

Abstract:

Helium from linear and tokamak-like plasma devices has been shown to cause "fuzz" to form on the surface of tungsten and other metallic surfaces after only a few hours of plasma exposure. The mechanisms involved occur on multiple length and time scales, and understanding all of those details is the subject of significant multi-scale modeling. This study examines the initial stages of fuzz formation by performing molecular dynamics simulations on both single-crystal and bicrystalline models of tungsten for times on the order of 0.5 to 2 microseconds. These simulations employ state-of-the-art many-body potential energy models and implantation depth distributions constructed from implantation simulations of surfaces with different crystal orientations. The large-scale atomistic simulations, which involve the simulation of up to 20 million atoms for up to 18 months or more, reveal the specifics of bubble formation and growth that drive surface evolution via the sequence of helium implantation, diffusion of helium to form bubbles, growth of bubbles and subsequent formation of tungsten self-interstitial atoms, organization of those self-interstitials into prismatic dislocation loops, annihilation of those prismatic loops at surfaces and grain boundaries, and finally, bubble rupture. The orientation of the surface and the presence of grain boundaries and pre-existing dislocations have pronounced effects on the pattern and quantity of helium retained and the resulting surface features. In particular, grain boundaries serve as sinks of helium, dramatically changing the distribution of bubbles in the vicinity of a grain boundary. These results provide important information about the morphological evolution of plasma-facing surfaces and provide important benchmarks for models of nuclear fusion reactor materials.

427719 Benchmarking of Heep with G4-Econs Program for Nuclear Hydrogen Production Plant

Abstract:

Hydrogen production with nuclear energy has been considered as one of the most economical hydrogen production methods, because it enables us to produce massive hydrogen at a relatively low price without environment pollution.

HEEP(Hydrogen Economy Evaluation Program) and G4-ECONS(Excel based Calculation Of Nuclear Systems) are two popular software which are useful to evaluate the economy of the hydrogen production system. HEEP is a simple and easy modeling program which can calculate cost of hydrogen production, transportation and storage by simply providing input data such as total capital cost, initial fuel load cost, annual fuel load cost, O&M cost and decommissioning cost as a percentage of total capital cost.

G4-ECONS is a EXCEL-based modeling program which can calculate the Levelized Unit Energy Cost(LUEC) from the reactor plant and the Levelized Unit Hydrogen Cost(LUHC) from the hydrogen production plant. The LUEC is the unit cost of electricity or thermal energy available from the reactor. The heat and electricity energy are related by the thermodynamic efficiency of the reactor plant. G4-ECONS can calculate cost of hydrogen production by providing input data such as annual capital cost(total capital cost times capital charge factor), annual fuel cost, annual O&M cost and annual decommissioning cost.

In the paper, the LUHC calculated with G4-ECONS program were compared with that of HEEP for the benchmarking. The result indicated LUHC calculated with G4-ECONS program is about 20% higher than that of with HEEP. This difference is believed to be from inherent difference dealing with financial parameters between two programs. The parametric study was done for the major financial factors that would affect the hydrogen production cost and it indicated that the inflation rate, discount rate and depreciation period have relatively significant influence on the LUHC.

422882 Efficiency and Design of High-Temperature Solar-Thermal Reactors

Abstract:

Efficiency and Design of High-temperature Solar-thermal Reactors

Arto J. Groehn, Allan Lewandowski, Ronggui Yang* and Alan W. Weimer

Department of Chemical and Biological Engineering, University of Colorado, Boulder, CO 80303, United States.

*Department of Mechanical Engineering, University of Colorado, Boulder, CO 80309, United States.

The goal of this work is to facilitate the use of concentrated solar power to drive chemical reactions for such applications as carbothermal reduction and water splitting. A computational model interfacing radiative, convective and conductive heat transfer was employed to investigate optimal reactor design and scale-up.

Typically employed radiation models can be classified into Monte Carlo (MC) and finite volume (FV) based approaches. MC models are often employed to solve radiation in static systems where effects of scattering, absorption, emission as well as anisotropy caused by particles, gases or varying material properties can be neglected. In contrast, these effects are readily accounted for in a FV-based discrete ordinates (DO) model (Chui & Raithby, 1993). Its accuracy, however, is limited by the number of employed discrete solid angles as well as computational cells which may result in ray effect and false scattering errors, respectively (Martinek & Weimer, 2013).

To reduce computational cost without compromising accuracy, previously a hybrid MC-FV radiation modeling scheme was suggested for a solar-thermal process (Martinek & Weimer, 2013). In this approach the incoming specular reflections were solved with a MC code while a FV model was used for the diffusive thermal re-radiation away from the reactor walls. Here, this hybrid concept is extended to allow MC-FV coupling at an arbitrary interface thus enabling simulation of complex physical mechanisms (e.g. temperature dependent emissivity) and their interactions inside the reactor domain.

The accuracy and optimal discretization level of the present approach was evaluated by comparing its predictions with those of a well-established and validated MC ray-tracing code SolTrace (Wendelin, 2003). Simulations with the validated model were conducted to determine flux profiles and thermal efficiencies of different heliostat field, secondary concentrator and receiver designs for large-scale (gigawatt) chemical reactors.

References

Chui, E. and G. Raithby, Computation of radiant heat transfer on a nonorthogonal mesh using the finite-volume method. Numerical Heat Transfer, 1993. 23(3): p. 269-288.

Martinek, J. and A.W. Weimer, Evaluation of finite volume solutions for radiative heat transfer in a closed cavity solar receiver for high temperature solar thermal processes. International Journal of Heat and Mass Transfer, 2013. 58(1): p. 585-596.

Wendelin, T. SolTRACE: a new optical modeling tool for concentrating solar optics. in ASME 2003 International Solar Energy Conference. 2003. American Society of Mechanical Engineers.

423075 Impact of Reduction of Flowing Particles on System Efficiency for Solar Thermochemical Hydrogen Production

Abstract:

Overall system efficiency for solar thermochemical hydrogen production is critically important. Here, an analysis of different reduction methods is examined using cerium oxide as a well-studied example material. These calculations include gas heating recuperation and thermodynamic separation work requirements, as well as a discussion of the relative sensitivity of each of these factors. The efficiency calculation uses thermodynamic equilibrium states of the active material to determine the reactivity and production capacity. Calculations are performed for flowing particles through a two-step solid oxide reduction and oxidation cycle. Spray dried particles for use in such a process were synthesized and characterized.

(200c) Measurement of Oxygen Release from Redox Samples for Solar Fuel Production

Lapp, J. L., University of Minnesota
Rieping, R.,
Roeb, M.,
Sattler, C.,
Originally presented on: 11/9/2015 15:57:00 - 16:18:00

The characterization of reduction-oxidation chemistry is important for the ongoing development of thermochemical cycles for solar fuel production. Materials like cerium oxide and perovskites have been studied in depth for their fuel production potential and mechanical stability during temperature cycles in the range of 1000 to 1500 degrees C. Accurate characterization of these materials is dependent on measurement of the gaseous products. A particular challenge is the measurement of oxygen release from samples. In the current study, a thorough investigation is performed of several options for measuring oxygen release rates due to high temperature cycling of solid metal oxides. An infrared furnace capable of extreme heating rates is used to chemically cycle solid redox-active metal oxides. Oxygen sensors located in the high temperature reaction tube and after gas cooling are compared to measurements from a mass spectrometer. Experimental data are given for qualification of the system using known gas compositions and for actual chemical cycling experiments. The potential errors due to time delays and diffusion effects are quantified. The resulting implications for future measurement systems are described.

433005 CO2 Reforming of CH4 with an Isothermal Redox Membrane Reactor

Abstract:

The continuous production of CO and H_2 via dry reforming of CH_4 is demonstrated with an isothermal redox membrane in absence of additional catalysts. The reactor technology realizes the continuous splitting of CO_2 to CO on the inner side of a tubular ceramic membrane and the partial oxidation of CH_4 with the lattice oxygen to syngas on the outer side. $La_{0.6}Sr_{0.4}Co_{0.2}Fe_{0.8}O_{3-\delta}$ (LSCF) membranes evaluated at 840-1030°C yielded up to 1.27 μmol CO from CO_2 g^{-1} LSCF s^{-1}, 3.77 μmol H_2 from CH_4 g^{-1} s^{-1}, and CO from CH_4 at approximately the same rate as CO from CO_2. To understand how CH_4 activation limits CO_2 conversion with LSCF, we compute the free energy of the oxygen vacancy formation for $La_{0.5}Sr_{0.5}B_{0.5}B'_{0.5}O_{3-\delta}$ (B, B' = Mn, Fe, Co, Cu) via electronic structure theory. Furthermore, we employ density functional theory to show how CO_2 conversion can be increased with advanced redox materials such as $La_{0.5}Sr_{0.5}MnO_{3-\delta}$ and $La_{0.5}Sr_{0.5}Mn_{0.5}Co_{0.5}O_{3-\delta}$ with lower oxygen vacancy stabilities. The developed principles may be useful for the rational design of advanced redox materials for solar-driven processes.

435513 Solar Thermochemical CO2 Capture Via Calcium Oxide Looping: Thermal Transport Modeling and Solar Reactor Design

Abstract:

The thermochemical cycle of calcium oxide looping is well suited for use with concentrated solar irradiation for carbon dioxide (CO_2) capture. The cycle consists of the reversible calcination–carbonation heterogeneous solid–gas reaction pair. CO_2 is chemically absorbed from a dilute source by a calcium oxide (CaO) sorbent to form solid calcium carbonate ($CaCO_3$) in the exothermic, non-solar carbonation step:

$$CaO + CO_2 \rightarrow CaCO_3,\ Dh^0 = -178\ \mathrm{kJ\ mol^{-1}}$$

The sorbent is regenerated and a stream of concentrated $CO_2$is released in the endothermic, solar-driven calcination step:

$$CaCO_3 \rightarrow CaO + CO_2,\ Dh^0 = 178\ \mathrm{kJ\ mol^{-1}}$$

Length scales involved in the realization of solar thermochemical technologies such as calcium oxide looping carbon dioxide capture typically range from 10^1–10^3 m at the solar plant level, to 10^{-1}–10^1 m at the reactor level, to 10^{-2}–10^{-1} m at the reactor components level, and to 10^{-8}–10^{-2}m at the component features level. Thermal transport models developed at the distinct length scales inform the design of the materials, reactors, and the overall process.

At the intra-particle level, a transient thermal transport model is used. It predicts heat and mass transfer coupled to chemical kinetics inside sorbent particles undergoing cycling. Both the solid and fluid phases are considered in the model. The model is used to investigate effects of the physical parameters such as particle size, and reactor operational parameters such as incident solar irradiation and carbon dioxide concentration. At the reactor component level, heat and mass transfer coupled to chemical kinetics is investigated for a packed bed of sorbent particles using computational fluid dynamics. The packed-bed dimensions are varied to study their effect on pressure drop, temperature distribution, and heat transfer in the packed-bed. On the reactor level, radiative exchange within the reactor cavity is analyzed using the net radiation method. The radiative exhange model is coupled to the heat and mass transfer model of the packed-bed to investigate the thermal transport phenomena occuring in the reactor.

A 1 kW_{th} reactor concept is proposed for solar CaO looping. The concept consists of a beam-up oriented, dual-cavity reactor wherein a receiver cavity is surrounded by an annular packed bed of sorbent particles. During carbonation, a CO_2-containing gas flows through the packed-bed reaction zone. The CO_2 chemically reacts with the sorbent particles, forming $CaCO_3$, while CO_2-depleted gas leaves the system. During calcination, concentrated solar energy enters an aperture located at the bottom of the reactor. The concentrated solar energy is captured by the inner cavity and transferred by conduction through a diathermal cavity wall to the packed-bed reaction zone, driving the calcination reaction. CO_2 liberated from the calcination reactor is removed from the reactor in a concentrated stream appropriate for storage or further use and processing. The reactor prototype design is informed by numerical thermal transport models at the intra-particle, packed-bed and reactor levels. The final reactor design is presented.

423221 Carbonation Kinetics of Sro By CO2 for Solar Thermochemical Energy Storage

Abstract:

Concentrated solar power (CSP) is one promising method of converting clean solar thermal energy into electricity which avoids the use of fossil fuels. Thermal energy storage (TES) in conjunction with CSP can increase the utilization of solar energy by enabling plant operators to generate electricity beyond normal on-sun hours. Thermochemical energy storage (TCES) is an emerging type of TES system based on a reversible reaction that offers higher energy density than latent or sensible energy storage. Thermochemical energy storage of on-sun thermal energy is achieved when a reactive system absorbs thermal energy and proceeds with a reversible chemical reaction. In a time of off-sun power demand, the reverse reaction is then initiated and energy is released, thus recovering thermal energy for use in a power cycle. One such reactive system is the reversible carbonation/decomposition of $SrO/SrCO_3$, which occurs ca. 1200°C. Such high quality heat is suitable for high efficiency, combined cycle power generation, which has the potential to translate into more competitive solar electricity prices.

In order to accurately predict the rate of heat generation from a TCES system and improve the CO_2 capture capacity of SrO, it is important to understand the mechanism of strontium oxide carbonation and the kinetic expressions describing the reversible reaction, which has not been studied. To investigate this mechanism, SrO carbonation kinetics have been studied by means of thermogravimetric analysis (TGA). In order to better describe the reaction mechanism, the influence of intensive variables such as reaction temperature and CO_2 partial pressure were investigated. Carbonation was performed at several temperatures (900-1200°C) and partial pressures of CO_2 (0.1 to 1 bar). The effects of different initial mass (20-60 mg) and sample size (25 – 106 μm) has also been investigated.

The carbonation reaction progression can be divided into three stages: an initial induction period; a rapid kinetically-controlled carbonation stage; and finally a sluggish diffusion-controlled regime. Our research aims to present the kinetic model developed to explain the experimental data and compare the heat generation rates to the current state of the art TES systems.

422896 A Novel Reduced Pressure Hybrid Solar/Electric Reactor for Producing Magnesium Metal Via Carbothermal Reduction

Abstract:

A novel 10kW hybrid/solar reactor operating at a temperature up to 1650°C and pressure down to 1 mbar for magnesium metal production via carbothermal reduction reaction ($C + MgO \rightarrow CO + Mg$) is presented. The unique reactor design includes a pneumatic actuated shutter that moves the solar concentrator and replaces it with a section of refractory insulation in front of the aperture to block the thermal radiation from the inside of the heated chamber when solar radiation is not applied. During daytime concentrated solar power is used to drive the chemical reaction and substitute high peak cost electricity. During nighttime off-peak low cost electricity is used to enable a 24/7 operation. Compared to solar-only applications which are limited to daylight operation, a hybrid solar/electric concept increases the plant's capacity factor and minimizes the financial risk. Key R&D efforts are determining the feasibility of heating the reactor tube using concentrated sunlight in addition to direct electrical resistance as well as the efficient collection of Mg avoiding back-reaction of Mg(v) and CO to MgO and C.

Developments in the Integrated Flowsheet for Treating Hanford's Radioactive and Hazardous Liquid Waste

Stuart Arm, Julie Colby, Renee Davis, Vinh Nguyen, Rose Russell
(Washington River Protection Solutions)

AIChE 2015 Annual Meeting
November 10th, 2015

Contents

- River Protection Project Integrated Flowsheet Overview
- Direct Feed Low Activity Waste Integrated Flowsheet
  - Overview
  - Development Tools
- Direct Feed Low Activity Waste Integrated Flowsheet Development
  - Value Stream Analysis
  - Rapid Improvement Events
  - Specific work activities (Just-Do-It's)
- Concluding Remarks

2

RPP Integrated Flowsheet Overview

- Ensure all RPP facilities work harmoniously together to efficiently treat, immobilize and disposition the tank waste.
- Mature the RPP Integrated Flowsheet by identifying and addressing gaps and opportunities:
  - Waste Acceptance Criteria and feed qualification methods
    - Definition
    - Understand WAC technical bases to identify gaps and opportunities.
  - Ability to satisfy or project satisfying (using process models) WAC
    - Identify flowsheet model gaps and opportunities.
  - Ability to satisfy or project satisfying (using process models) throughput targets
  - Flowsheet optimization.
- Develop and implement flowsheet maturation plans to close gaps and realize opportunities.
- *Incorporates and enables the National Laboratory's role of providing continuity of scientific expertise and technical defensibility through laboratory integration across the elements of the RPP Mission Flowsheet.*
  - Pacific Northwest and Savannah River National Laboratory staff 'embedded' in the Integrated Flowsheet team.

3

RPP Integrated Flowsheet Overview

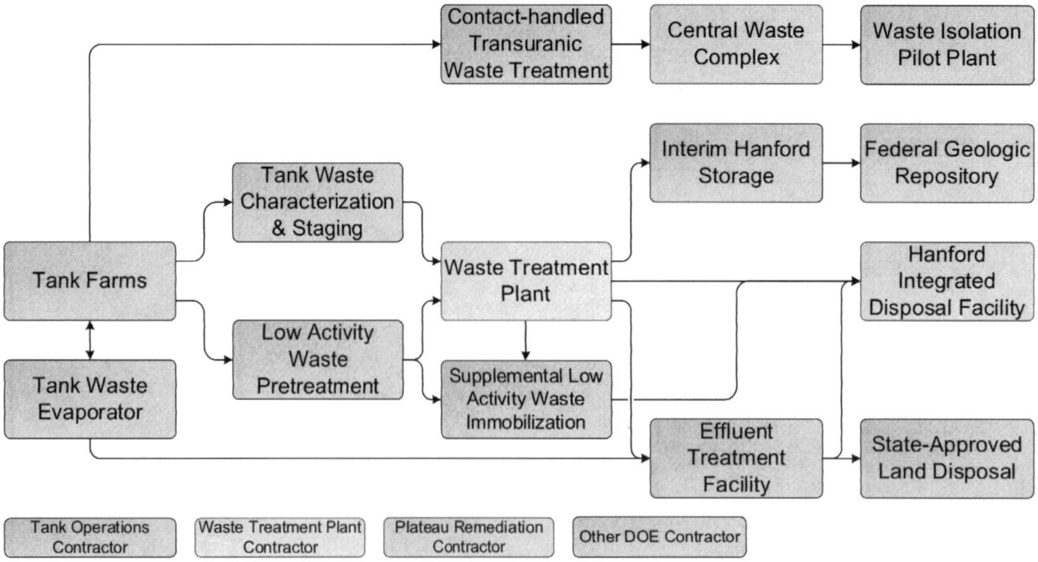

4

DFLAW Integrated Flowsheet

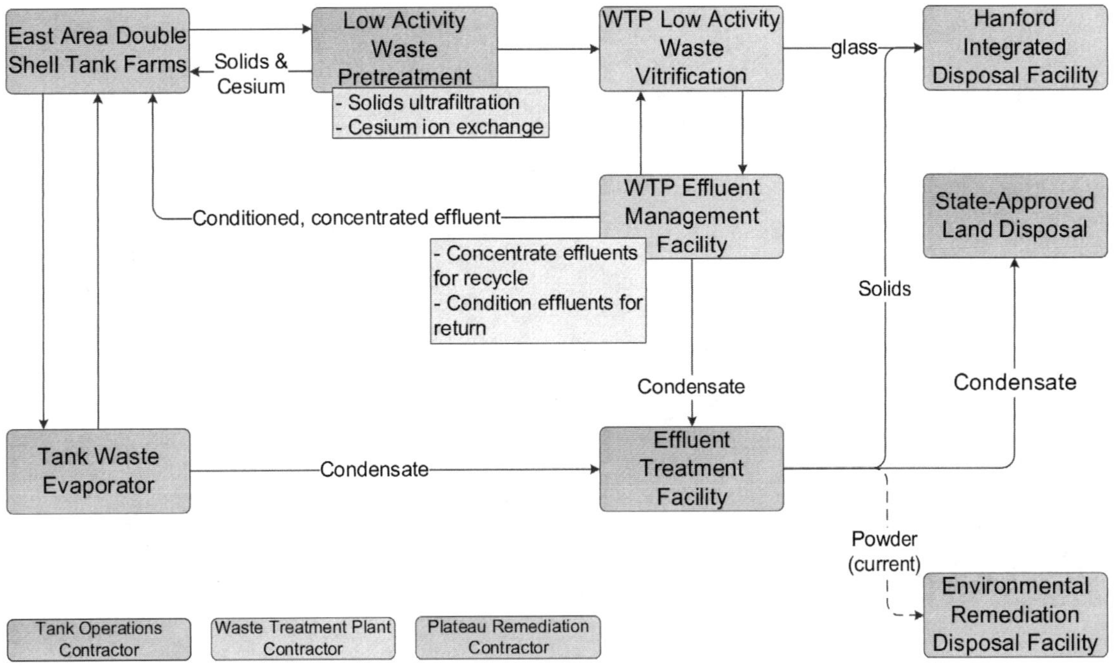

5

DFLAW Integrated Flowsheet Development Tools

- WRPS has introduced the Lean management system in a renewed effort to reduce waste and non-value added work.
  - Lean management is a system of continuous improvement that can be applied to any work process.
  - Simplifies and streamlines processes in the tank farms to improve performance.
  - The system looks at the flow of work and teaches workers how to identify and remove inefficiencies to improve the work process.
- DFLAW Value Stream Analysis
  - Week-long activity in which a team of experts was assembled to identify areas for improvement.
- Rapid Improvement Events
  - Focus improvement areas developed by expert teams over a week.
  - Actions identified.
  - Four events conducted: Cesium eluate return; DFLAW effluent volume & chemistry; DFLAW feed qualification; sodium feed concentration optimization.
- Specific work activities pursued by small teams and individuals targeted at areas for improvement over several weeks ('Just-do-it').

6

DFLAW Value Stream Analysis Event

EVENT INFORMATION	
Start Date	December 1, 2014
Duration	4 days
Sensei	Steve Brune
Process Owner	Renee Spires
Team Leader	Renee Spires
Facilitator	Mark Kemner
Team Members	Stuart Arm, Rose Russell, Murray Thorson, Jacob Reynolds, David Reinemann, Michael Stone, Matthew Fountain, Tom Nirider, Matt Mills, Brion Taki, Peter Benson, Tony Waldo, Ben Gallaher, Kayla Pierson, Dave Herman

7

DFLAW Value Stream Analysis Gap Analysis

Methods used to meet requirements are hard
- Labor intensive/administratively burdensome tank sampling & analysis
- Many long duration sample requirements

Waste Acceptance Program needs definition
- Lack of standard WAC
- No clear system of guidance for showing compliance

Processes do not have defined requirements

Too many requirements
- Organics
- Inorganics

Lack of Integration between Facilities

Lack of surge capacity between LAWPS and WTP LAW Vitrification Facility

Impact of volume is unknown
- Too much fresh water addition
- Too much eluate volume
- Recycle within LAW-vit vs. send to DSTs
- Difficult water management

Infrastructure Impacts
- Obtaining timely client approval (LAWPS)
- Large investments will be required
- Limitations of aged infrastructure (DSTs)

LAW Vit Throughput
- Impacts of halides, sodium concentration
- Waste loading

LAWPS Throughput
- Dilution to mitigate precipitation
- Difficult waste (supersaturated)

Impact of Chemistry is Unknown
- Recycle to DSTs
- Corrosive chemical accumulation
- Ability to handle high halide concentration

Long Lead Regulatory Processes
- NEPA EIS constrains disposal options/revisions needed
- IDF PA delays ILAW

8

DFLAW Value Stream Analysis Solution Approach

If we...	Then we...
Eliminate recycling to the Tank Farms	Uncouple Retrieval Operations and DF-LAW processing interdependency
Keep recycles within LAW-Vit	Eliminate the need for some 242-A Evaporator campaigns & generate major $$ savings
Reduce volume of cesium streams returned to DSTs	Reduces DST Space impacts & increases future HLW processing flexibility
Send ETF waste to ERDF, not IDF	Eliminate need for ETF grout facility
Perform process sampling & analysis at the beginning of the process	Can have continuous operations and dramatically reduce touch & flow time and cost

9

Cesium Eluate Volume Management

- Problem
  - The cesium eluate sent from LAWPS to the DST System reduces the available DST storage space and thereby could limit operational flexibility.
- Adding a plant wash tank to LAWPS enables segregation of most of the separated cesium into a single DST.
  - Increases LAWPS project costs but overall mission savings.
  - Considered adding the penetrations necessary to convert the Resin Dump Tank to dual purpose to receive plant wash.
    - Subsequent design changes now make this option inappropriate and the opportunity is no longer being pursued.

10

Effluent Return to DST System - Chemistry

- DFLAW effluents are substantially different from the tank waste.
  - Elevated concentrations of halides and sulfate and low nitrate concentration means pitting is the predominant corrosion mechanism.
  - Current DST System corrosion control criteria aimed at mitigating stress corrosion cracking.
- Flowsheet models and acceptance criteria currently using SRS corrosion control criteria developed for return of DWPF effluents on a preliminary basis.
  - SRS criteria qualified for lower concentrations of halides than projected for DFLAW.
    - Adjust total halides (F^- and Cl^-) concentration $\leq$ 0.05M and $[SO_4^{-2}] \leq 0.25M$
    - Verify effluent temperature $\leq 35^oC$ (or 95^oF)
    - Verify pH > 10.3.
    - Verify $[NO_3^-] < 1.0M$
    - Achieve minimum needed nitrite concentration by selecting the highest [NO2-] value from three correlations below:
      - $[NO_2^-] = 1.02*[NO_3^-]^{0.98}$
      - $[NO_2^-] = 166.4*[X^-]^{1.34}$ where $[X^-]$ is total halides (F^- and Cl^-) concentration and is $\leq$ 0.05M.
      - $[NO_2^-] = 1.09*[SO_4^{-2}]^{0.835}$ this equation is only applied if $[SO_4^{-2}]$ is $\leq$ 0.25M
- Ongoing testing is expanding the range of applicability (to 0.1M total halides) for the SRS corrosion control criteria to avoid the initial dilution.
  - Initial results show range can be extended to 0.2M and nitrite concentration of 3M mitigates corrosion across the range.

11

Effluent Return to DST System - Volume

- The halide, sulfate and sodium concentrations in the melter feed all can limit the waste loading in glass.
  - Recycling effluents directly to the melter will increase the concentrations of sulfate and halides above that otherwise present with no recycle.
    - Immediate reduction in waste loading and increase in immobilized LAW quantity.
  - Returning effluents to the DST System require addition of sodium nitrite.
    - Make more immobilized LAW later (post DFLAW).
    - Competes for DST space with retrieved SST waste

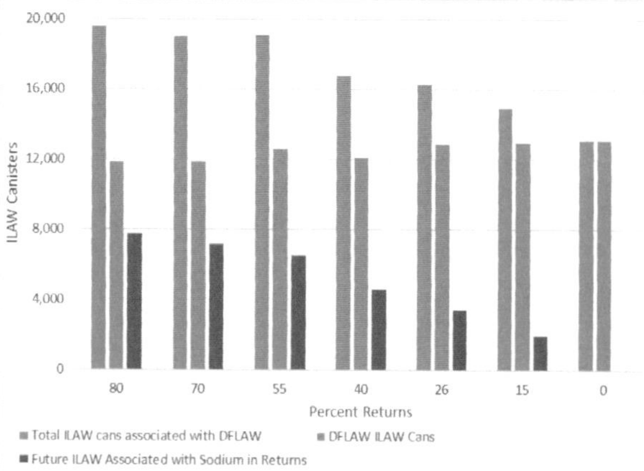

12

Optimize Feed Sodium Concentration

- WTP LAW feed basis is different from LAWPS than WTP Pretreatment

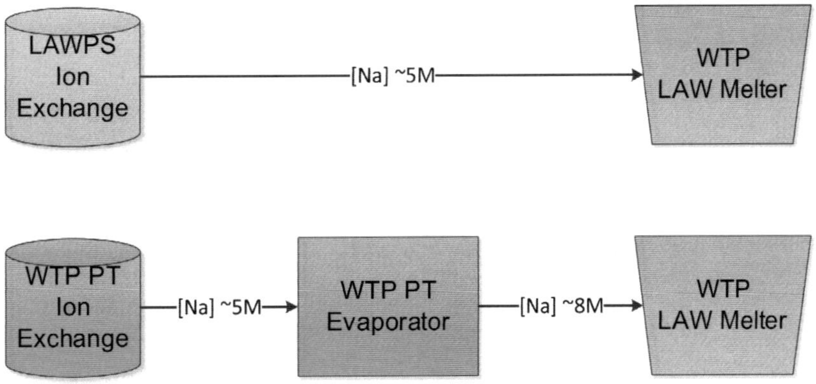

13

Optimize Feed Sodium Concentration

- The lower sodium concentration potential impacts on WTP LAW include:
  - the potential for decreased waste Na throughput of the melter.
  - decreased cycle time requirements for LAW feed and offgas condensate tanks.
  - increased carryover of particulate into the LAW offgas system.
  - higher velocities in the offgas system.
  - increased condensation load on offgas system.
- Solutions include future tests and evaluations:
  - Evaluate LAWPS process performance at higher [Na].
  - Research line flushing basis and optimize.
  - Melter feed rheology requirements.

14

Integrated Feed Qualification

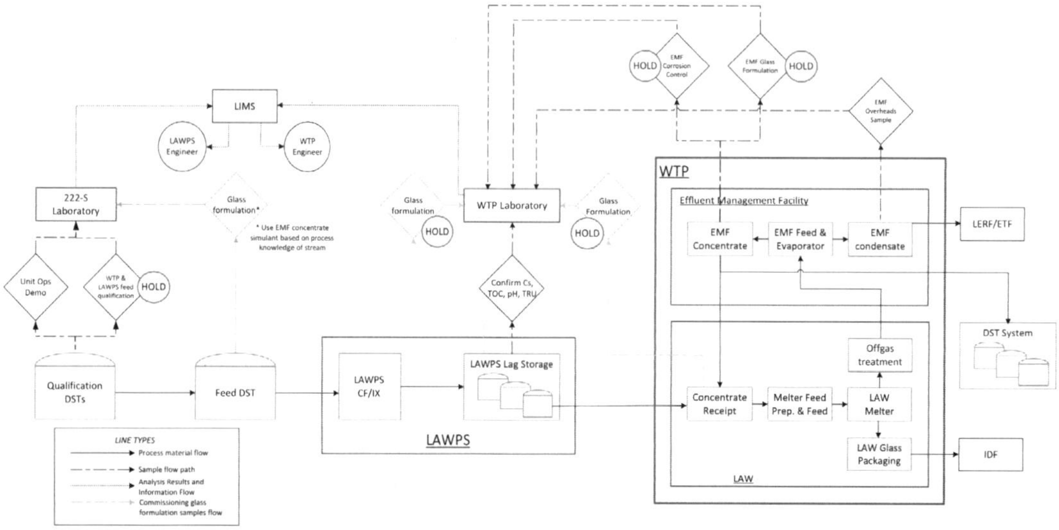

15

Concluding Remarks

- Integrated flowsheet for directly feeding treated LAW to WTP is established as a component of the complete RPP Integrated Flowsheet.
- Optimization of the flowsheet is underway and this year used the Lean management system as a tool to:
  - Identify areas for improvement.
  - Develop those areas for improvement.
    - Effluent management (cesium product, WTP effluent chemistry and volume).
    - Feed composition (sodium concentration).
    - Integrated feed qualification.
- Next year:
  - Further optimize and develop the integrated feed qualification concept.
  - Refine the volume of effluent returned to the DST System.
  - Address lack of technical bases for line flushing and slurry transport.
  - Quantify uncertainties in flowsheet model inputs and perform sensitivity analyses.

16

Questions?

416757 Electrochemical Recovery of Fission Platinoids in 1-Butyl-1methylpyrrolidinium Dicyanamide Ionic Liquid

Abstract:

Electrochemical Recovery of Fission Platinoids in 1-Butyl-1methylpyrrolidinium Dicyanamide Ionic Liquid

Sujan Shrestha[a], Miyan Nagib[a], Elizabeth J. Biddinger[a]* [a]Department of Chemical Engineering, City College of New York, 140th Street and Convent Avenue, New York, New York 10031 *Corresponding Author ebiddinger@che.ccny.cuny.edu

Ionic liquids (ILs) have been proposed as a safer solvent than volatile organic compounds (VOCs) for extraction of metals from aqueous sludge of spent nuclear fuel (SNF) due to its low flammability and thermal stability (1-3). After extraction, the metals can be judiciously recovered through electrodeposition because of the metal complexing ability, moderate-to-high conductivity, and wide electrochemical window of the ILs (4). Therefore, studying electrodeposition of metals in ILs is of critical importance for the down-stream processing in the IL-based nuclear waste management processes.

In this work electrodeposition of non-radioactive counterparts of fission platinoids (Pd, Rh and Ru) in ILs was investigated in 1-butyl-1-methylpyrrolidinium dicyanamide ([Bmpyrr][DCA]). Besides being important for nuclear waste management, fission platinoids have high economic importance (5). Among the fission platinoids, Pd is the most utilizable because of its low radioactivity compared to that of Rh and Ru (5).

It was expected that the thermodynamics and kinetics of the Pd change in the presence of Rh and Ru (6). So, we studied electrochemical behavior of Pd, Rh, and Ru separately and together in [Bmpyrr][DCA] using cyclic voltammetry. Electrodeposition was carried out on Ni substrate separately and from their mixture. Pd was found to be more soluble than Rh and Ru in [Bmpyrr][DCA]. Pd could also be easily reduced on Ni substrate within the cathodic limit of [Bmpyrr][DCA] forming a thin film of different morphologies dependent upon the potential applied. How the presence of Rh and Ru affect the electrodeposition behavior of Pd will be discussed in this presentation.

References

1. S. Ha, R. Menchavez and Y.-M. Koo, *Korean Journal of Chemical Engineering*, **27**, 1360 (2010).

2. K. Venkatesan, T. Srinivasan and P. V. Rao, *Journal of Nuclear & Radiochemical Science*, **10**, R1 (2009).

3. X. Sun, H. Luo and S. Dai, *Chemical Reviews*, **112**, 2100 (2011).

4. A. P. Abbott and K. J. McKenzie, *Physical Chemistry Chemical Physics*, **8**, 4265 (2006).

5. Z. Kolarik and E. V. Renard, *Platinum Metals Review*, **49**, 79 (2005).

6. M. Jayakumar, K. A. Venkatesan, T. G. Srinivasan and P. R. V. Rao, *Desalination and Water Treatment*, **38**, 73 (2012).

418499 Electroytic Reduction of Zirconium Oxide in Molten LiCl-Li2O to Study Mass Transport of Li2O

Abstract:

Direct electrolytic reduction in molten chloride salt has been demonstrated for a number of different metal oxides. When using an electrolyte consisting of LiCl and Li_2O, minimization of lithium oxide retention in the reduced product is important for process efficiency and product quality. In this study, zirconium oxide reduction in this salt has been experimentally investigated. Cyclic voltammetry with a basket containing zirconium oxide as the working electrode was used to assess the relative importance of lithium oxide electrowinning versus direct electrolytic reduction of the zirconium oxide. It is theorized that a high rate of direct ZrO_2 reduction coupled with limited mass transport of Li_2O will result in excessive Li_2O buildup in the cathode basket. This effect was tested via electrolytic reduction experiments in which basket rotation and periodically interrupted cell current were used to minimize accumulation of Li_2O in the cathode basket. To provide data for future modeling, the solubility of metallic lithium was measured as a function of lithium oxide concentration in the lithium chloride. The reported experiments were performed in an argon atmosphere glove box, with low oxygen and moisture concentrations. All experiments were run at a temperature of 650°C. Electrode potentials were measured and controlled versus a nickel/nickel oxide reference electrode. The anode used for these experiments was made of platinum wire and was purged with argon to enable continuous measurement of oxygen gas produced from the reaction.

15AIChE
Annual Meeting, Salt Lake City, UT

Adsorption of radioactive I_2 and water from spent fuel reprocessing off-gases by reduced silver mordenite

Yue Nan[1], David W. DePaoli[2] and Lawrence L. Tavlarides[1]

[1] Department of Biomedical and Chemical Engineering, Syracuse University
[2] Nuclear Science and Technology Division, Oak Ridge National Laboratory

November 10, 2015

NEUP Nuclear Energy University Programs
U.S. Department of Energy

This work is supported by the Office of Nuclear Energy, US Department of Energy, Standard Research Contracts: NE0008275.

College of Engineering and Computer Science
SYRACUSE UNIVERSITY

ADSORPTION MODELING AND VERIFICATION

Single-layer Adsorption/Desorption

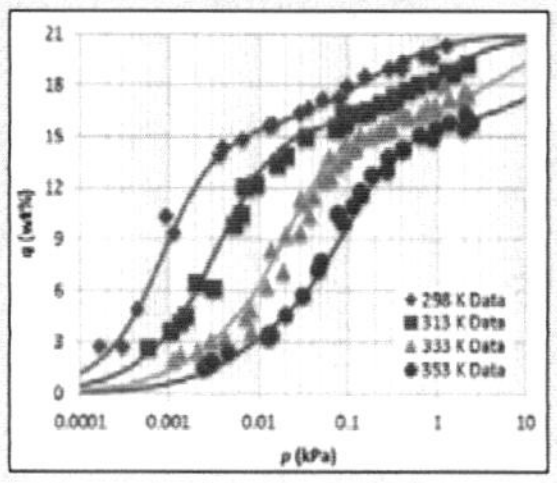

Isotherm Model

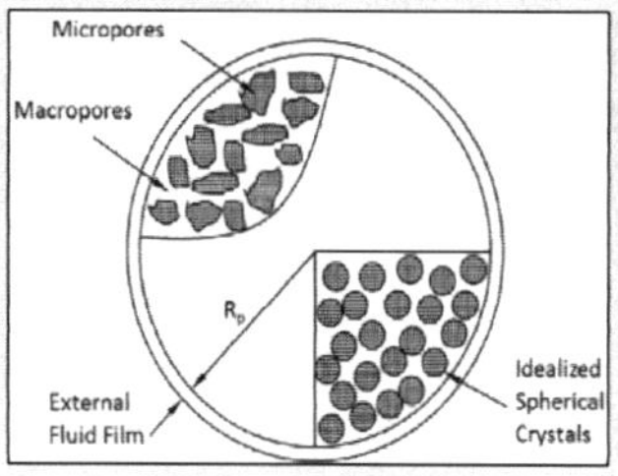

Diffusion Kinetics

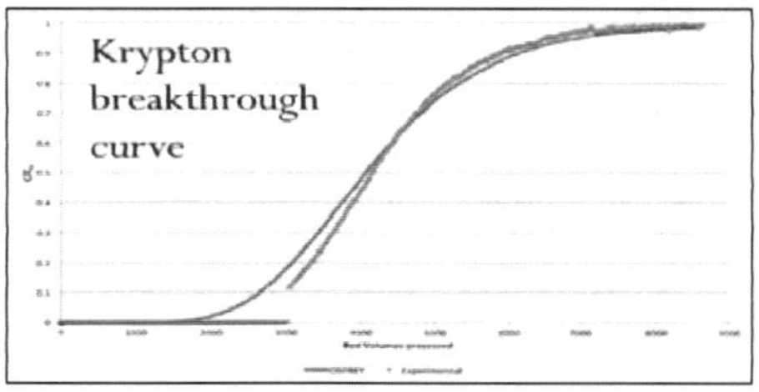

The OSPREY Adsorption Unit Operation Model*

Deep-bed Adsorption/Desorption

*V.J. Rutledge, OSPREY Model, INL/EXT-13-28150, 2013.
R. Lin,, A. Ladshaw, Y. Nan, J. Liu, S. Yiacoumi, C. Tsouris, D. W. DePaoli, and L. L. Tavlarides. I&EC 54, no. 42 (2015): 10442-10448.

College of Engineering & Computer Science
SYRACUSE UNIVERSITY

PART I: IODINE ADSORPTION

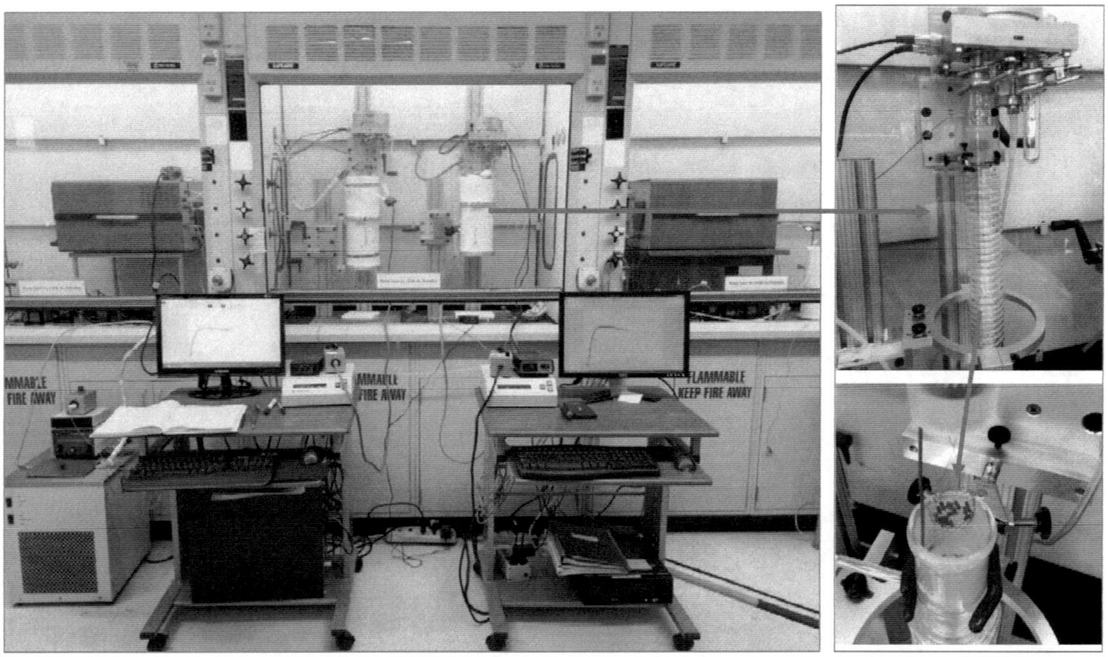

IODINE CAPTURE ON REDUCED SILVER MORDENITE

- ❑ The reduced silver mordenite (Ag^oZ) has better iodine adsorption capacity than the unreduced form(AgZ)
- ❑ Ag^+ ions are reduced and then migrate from pores to the external mordenite crystal surface, aggregating into nanoparticles
- ❑ Iodine is captured by forming both γ-AgI and α-AgI

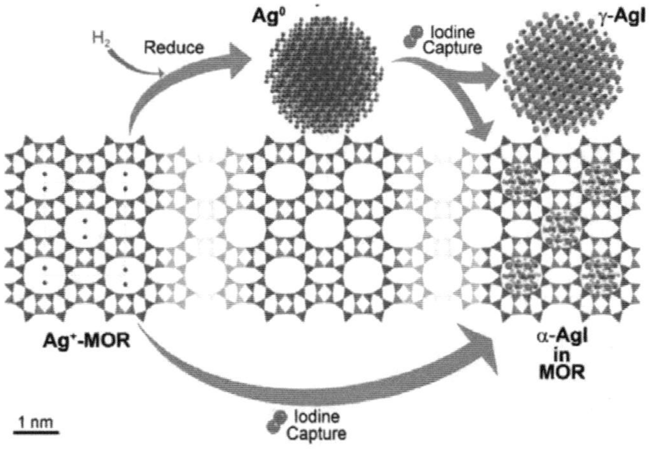

Chapman, K. W., Chupas, P. J., & Nenoff, T. M. (2010). Journal of the American Chemical Society, 132(26), 8897-8899.

EFFECTS OF REDUCTION CONDITIONS

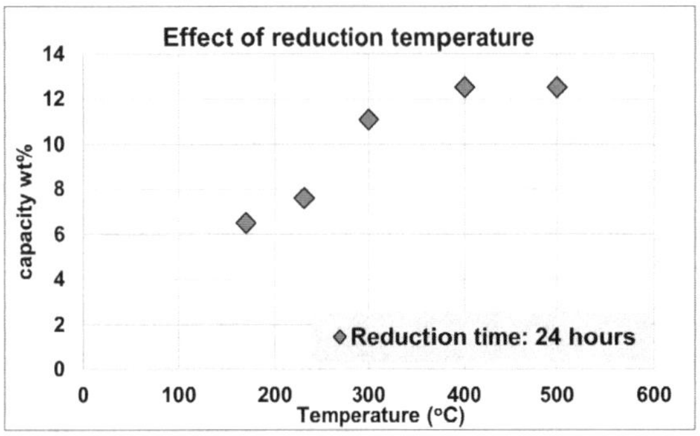

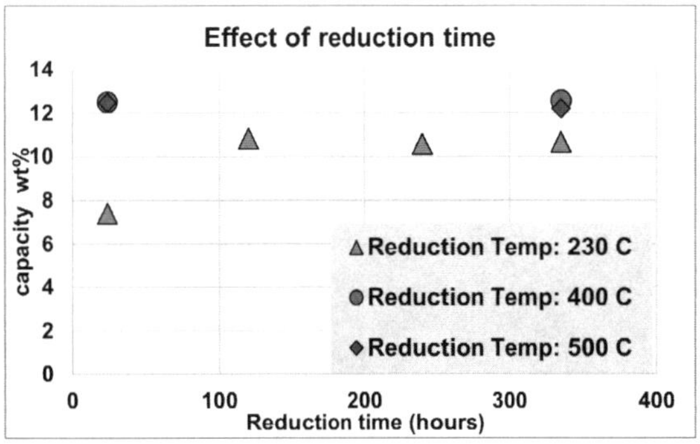

- Commercial AgZ :
  - IONEX-Type Ag-900
  - Size: 1.8 mm
  - Ag content: 11.9 wt%
  - Theoretical I_2 loading:14 wt%
  - Density: 3.06 g/ml
  - Porosity: 0.38

- Optimal reduction conditions:
  - Temperature: 400°C
  - Time: 24 hours

FLOW DIAGRAM OF THE ADSORPTION SYSTEM

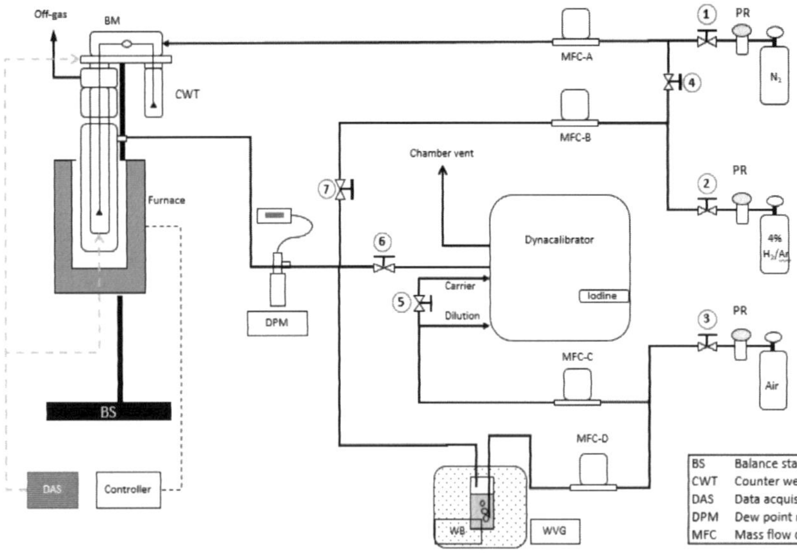

BS	Balance stand	MB	Micro balance
CWT	Counter weight tube	PR	Pressure regulator
DAS	Data acquisition system	WVG	Water vapor generator
DPM	Dew point meter	WB	Water bath
MFC	Mass flow controller		

UPTAKE CURVES OF IODINE ADSORPTION ON Ag^0Z

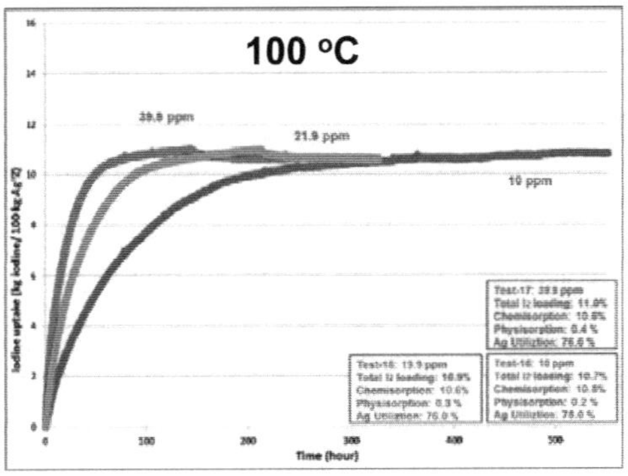

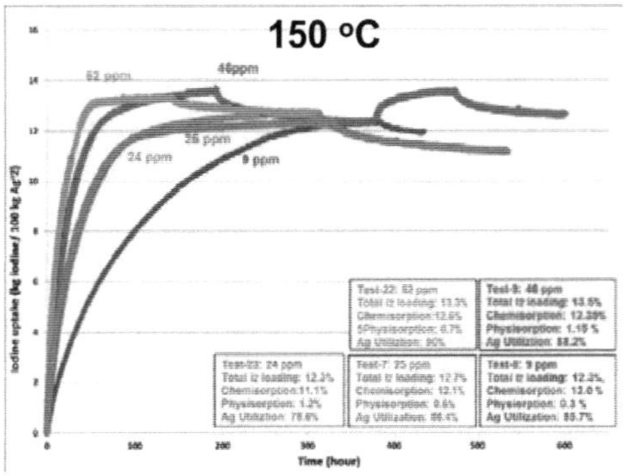

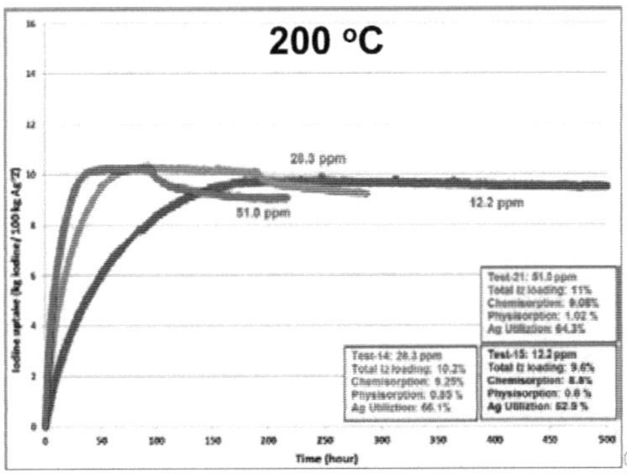

- Adsorption Temp.: 100,150, 200 °C
- I2 conc.: 9 – 52 ppmv
- Carrier gas dew point: - 70°C (dry air)
- Gas velocity : 1.1 m/min
- AgZ reduction conditions: 400°C, 24 hours
- Ag^oZ was first pre-equilibrated in dry air
- Desorption was performed with dry air
- Maximum I_2 loading: 13.5 wt%
- Maximum Ag utilization: 88.2 %

ISOTHERMS OF IODINE ADSORPTION ON Ag^0Z
TOTAL IODINE LOADING

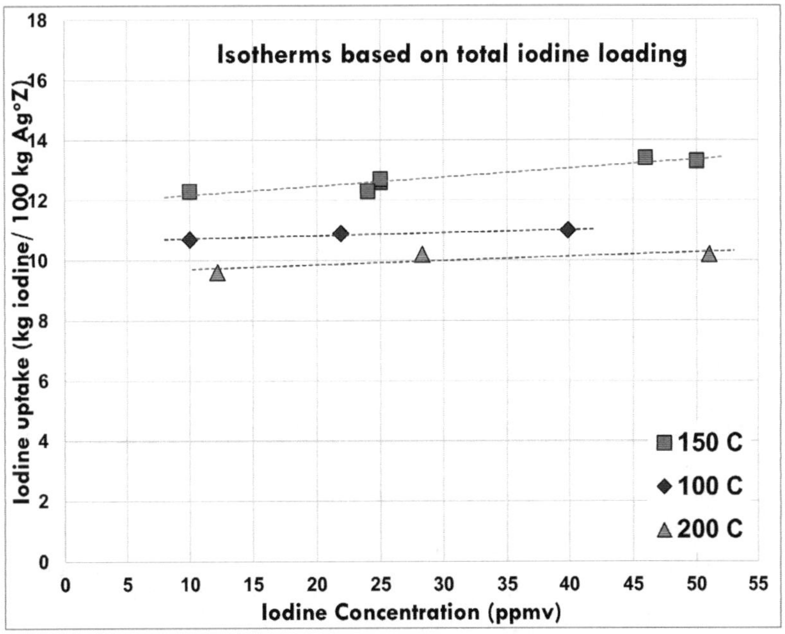

- ❑ Linear isotherms
- ❑ Optimal adsorption temperature:150°C
- ❑ Slope mostly due to the physisorption of iodine

ISOTHERMS OF IODINE ADSORPTION ON Ag^0Z
CHEMISORPTION

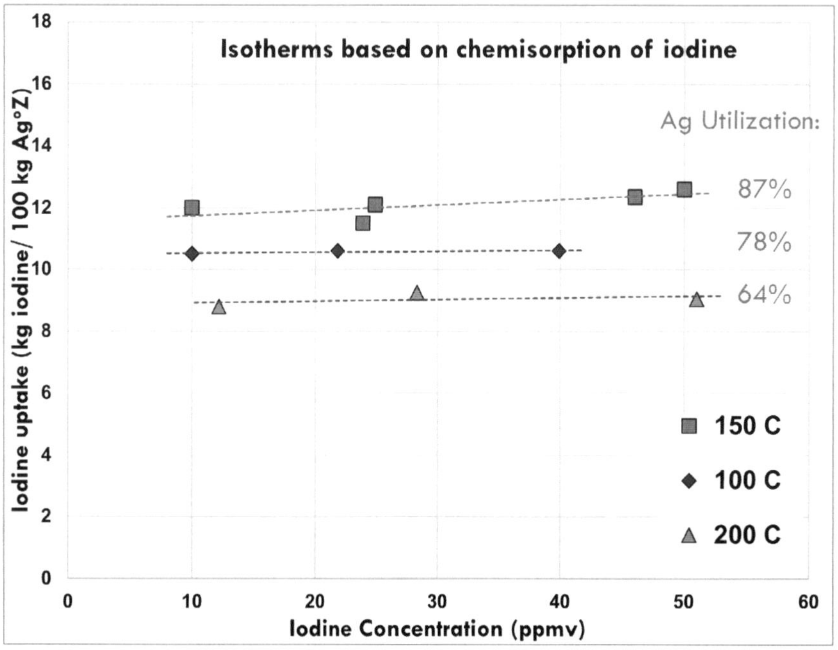

- Chemisorption of I_2 is almost constant at each temperature
- 150°C is the optimal temperature for chemisorption of I_2

ADSORPTION KINETICS MODEL

Shrinking-Core (SC) Model:

$$t = \frac{q}{q_e}\tau_1 + \left(1 + 2\left(1 - \frac{q}{q_e}\right) - 3\left(1 - \frac{q}{q_e}\right)^{2/3}\right)\tau_2 + \left(1 - \left(1 - \frac{q}{q_e}\right)^{1/3}\right)\tau_3$$

q – sorbate concentration in pellet
q_e – equilibrium sorbate concentration in pellet
r_a – radius of pellet
ρ_B – molar density of B
b – stoichiometric coefficient of B
k_g – film mass transfer coefficient
k_s – first order reaction constant
C_A – bulk gas-phase concentration
D_e – pore diffusivity

Assumption: $I_2 + 2Ag \rightarrow 2AgI$

$$A \xrightarrow{k_s} B$$

$$\tau_1 = \frac{r_a \rho_B}{3bk_g C_A}$$ Gas film control

$$\tau_2 = \frac{r_a^2 \rho_B}{6bD_e C_A}$$ Pore diffusion control

$$\tau_3 = \frac{r_a \rho_B}{bk_s C_A}$$ Reaction control

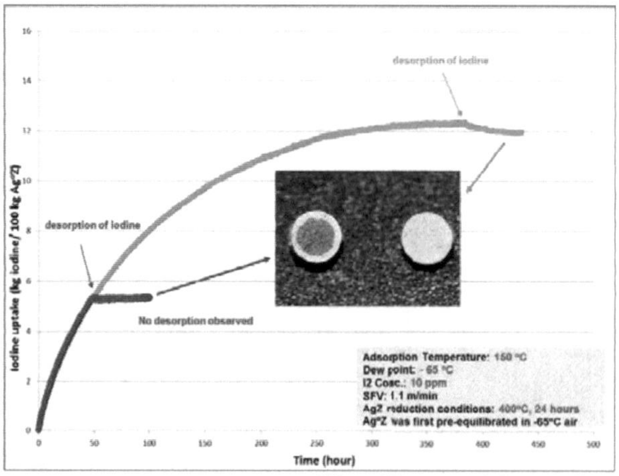

MODELING RESULTS

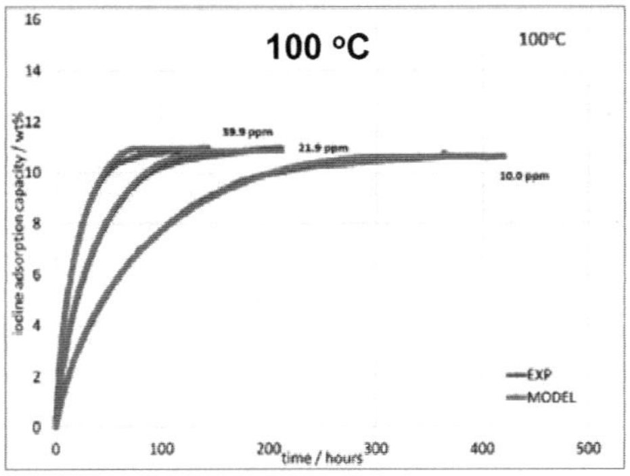

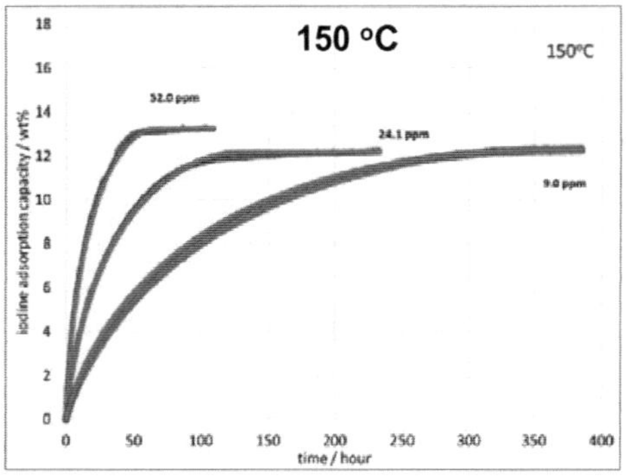

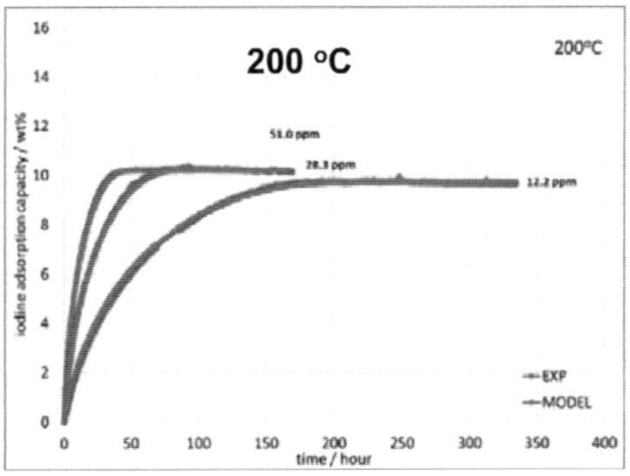

	AARD %
100 °C	1.42
150 °C	1.73
200 °C	1.26
Avg	1.47

- ❑ Excellent agreement between the model prediction and experimental data
- ❑ Average AARD is 1.47%

MODEL PARAMETERS

	kg cm/s	De cm^2/s	ks cm/s	AARD %
100 °C	6.49E+00	4.91E-3	0.132	1.42
150 °C	7.45E+00	5.84E-3	0.172	1.73
200 °C	8.42E+00	8.07E-3	0.196	1.26

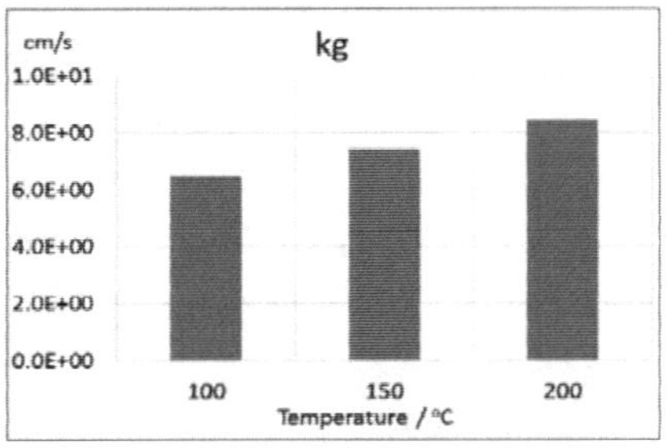

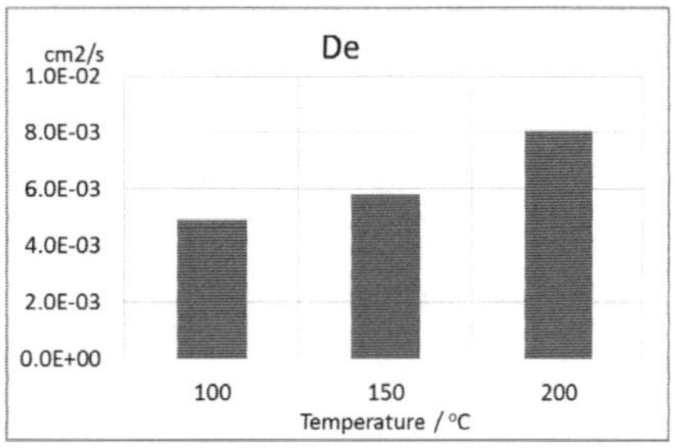

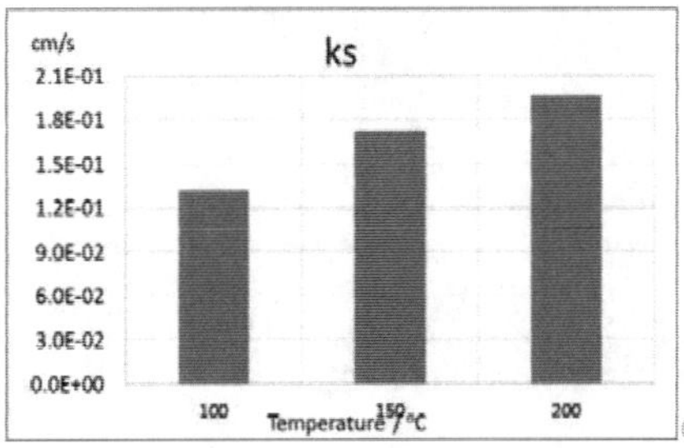

- ❑ The adsorption parameters increase with temperature
- ❑ High temperature favors uptake rate

*kg obtained by Renz-Marshall correlation
De and ks, obtained by least square fitting

IMPACT OF WATER IN THE GAS STREAM

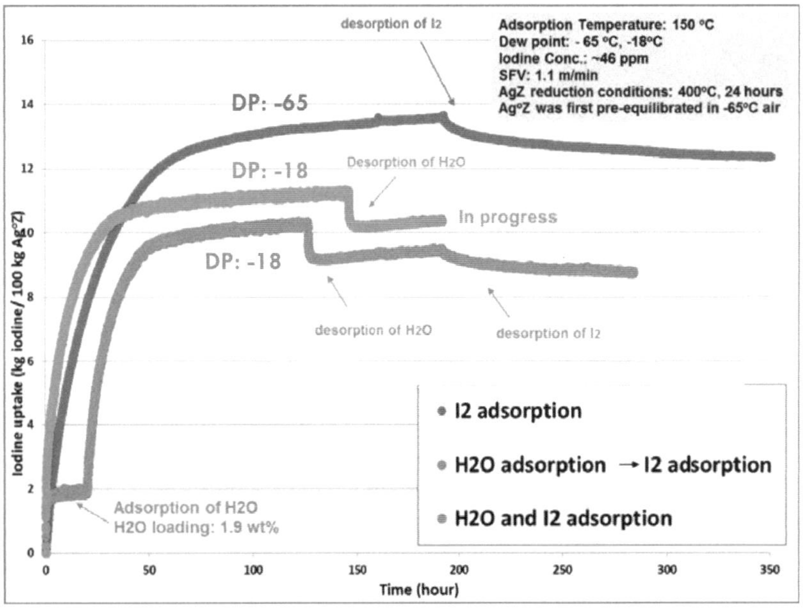

- ❑ Ag^oZ lost ~38% I_2 adsorption capacity after 1.9wt% of H_2O was loaded
- ❑ I_2 loading also decreased when simultaneously adsorbed with H_2O
- ❑ Co-adsorption models needed the describe the adsorption in real off-gases

COLLEGE OF ENGINEERING AND COMPUTER SCIENCE | SYRACUSE UNIVERSITY

College of Engineering & Computer Science
SYRACUSE UNIVERSITY

PART II: WATER ADSORPTION

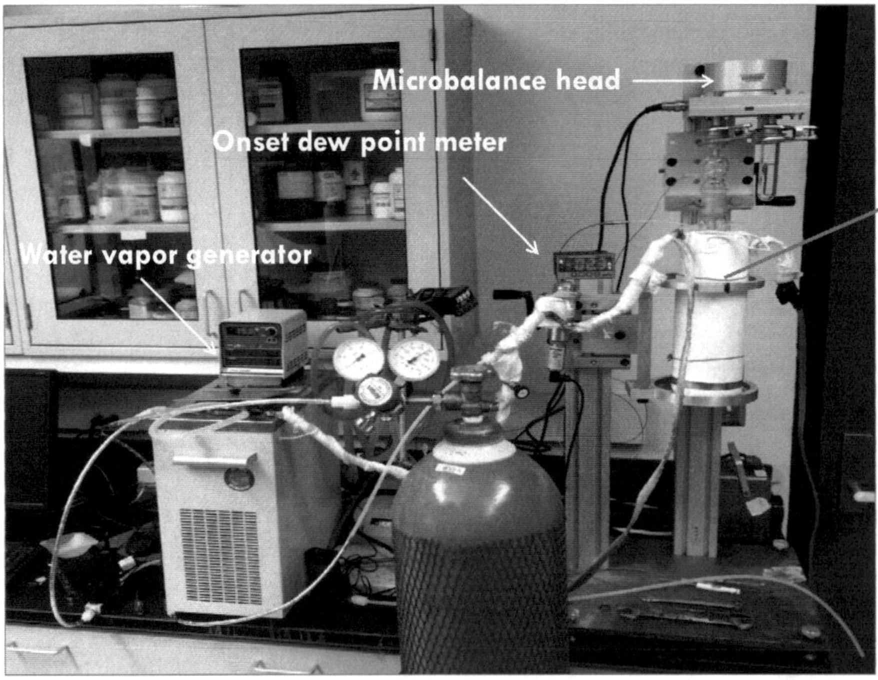

Ag^oZ sorbents

Single-layer Adsorption

- Equilibrium
- Kinetics

COLLEGE OF ENGINEERING AND COMPUTER SCIENCE | SYRACUSE UNIVERSITY

College of Engineering & Computer Science
SYRACUSE UNIVERSITY

WATER ADSORPTION UPTAKE CURVES AND ISOTHERMS

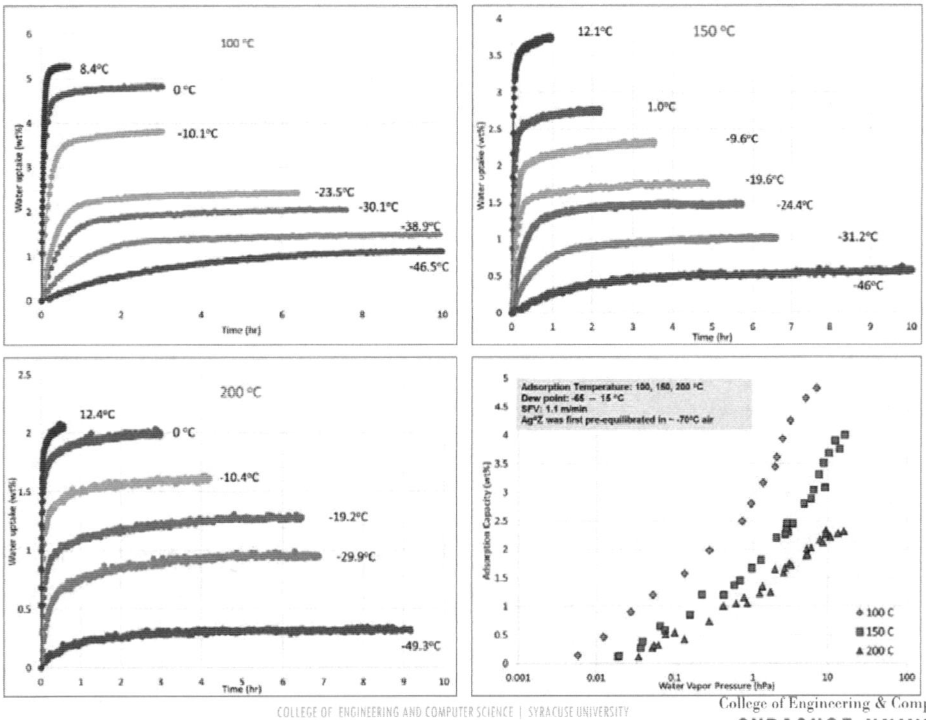

KINETIC MODELS

Linear Driving Force (LDF) Model

$$\frac{dq}{dt} = k\left(q_e - q\right)$$

$$\frac{1}{k} = \frac{R_p}{3k_f}\frac{q_e\rho_p}{C_b} + \frac{R_p^2}{15\varepsilon_p D_p}\frac{q_e\rho_p}{C_b}$$

Film resistance **Macropore resistance**

Shrinking-Core (SC) Model:

$$t_{tot} = \frac{q}{q_e}\tau_1 + \left(1 + 2(1-\frac{q}{q_e}) - 3(1-\frac{q}{q_e})^{2/3}\right)\tau_2$$

$$\tau_1 = \frac{R_p q_e}{3\varepsilon_p k_f C_b} \qquad \tau_2 = \frac{R_p^2 q_e \rho_p}{6\varepsilon_p D_p C_b}$$

Film control **Macropore control**

Kf obtained from: $Sh = \frac{k_f d_p}{D_{AB}} = 2 + 1.1\mathrm{Re}^{0.6}\, Sc^{1/3}$

MODELING RESULTS

LDF Model

	kf cm/s	Dp cm^2/s	AARD
100 °C	13.36	0.4079	2.59%
150 °C	15.45	0.3966	3.45%
200 °C	17.57	0.3238	4.10%
Avg			3.38%

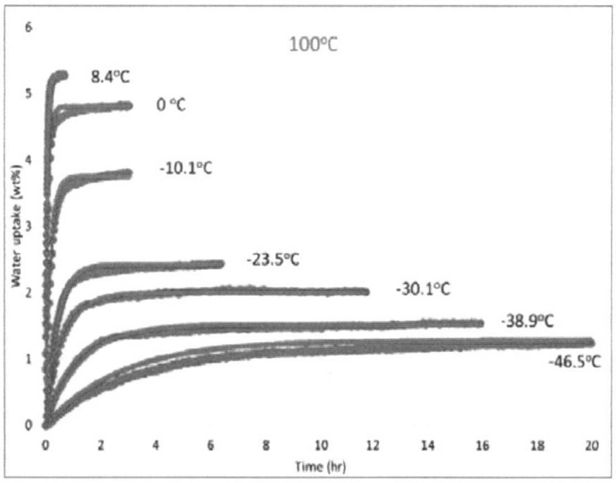

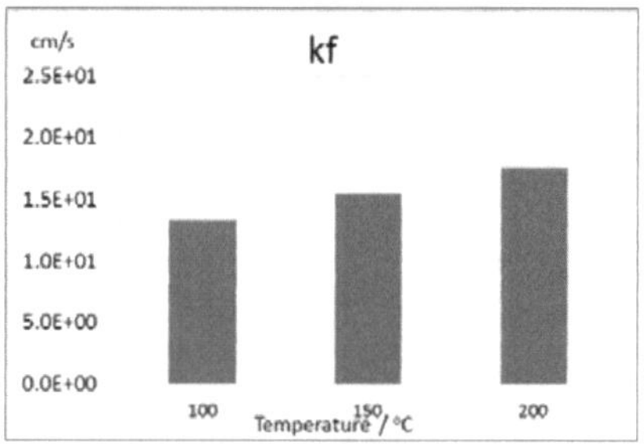

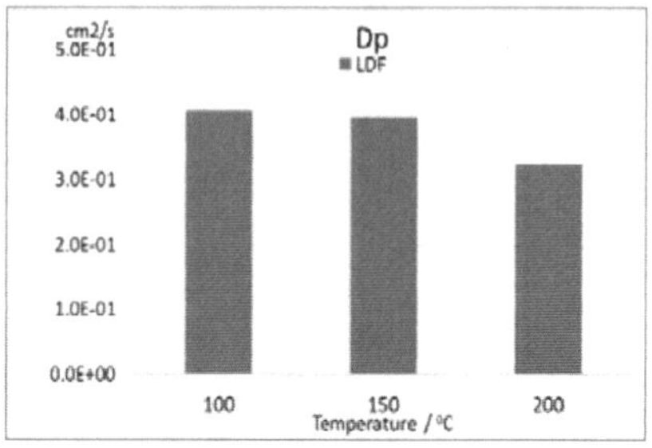

- ❑ LDF model is capable to predict the water adsorption on Ag°Z
- ❑ Average AARD is 3.38%.

MODELING RESULTS

SC Model

	kf cm/s	Dp cm^2/s	AARD
100 °C	13.36	0.2850	3.08%
150 °C	15.45	0.2673	3.71%
200 °C	17.57	0.1326	3.50%
Avg			3.43%

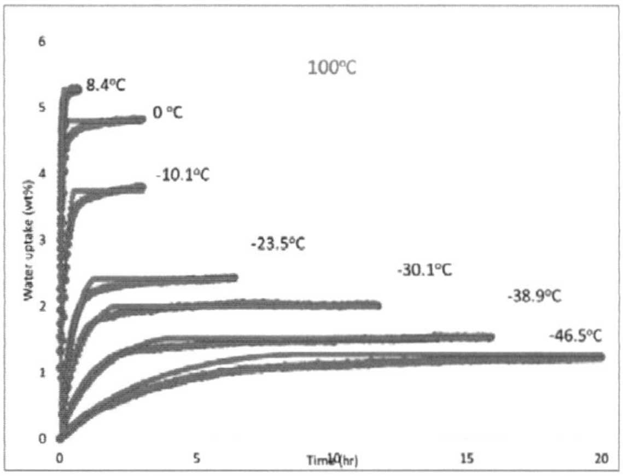

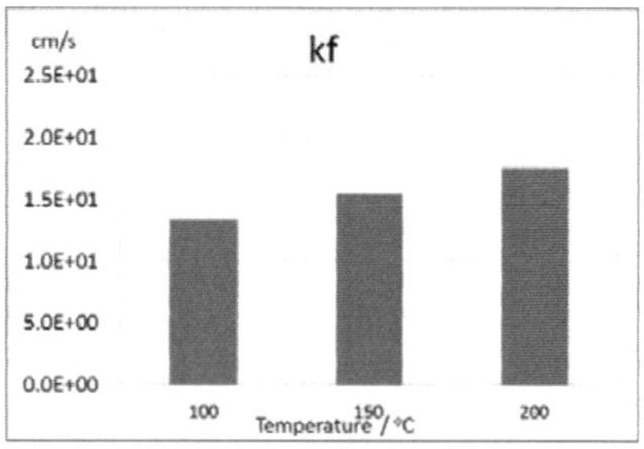

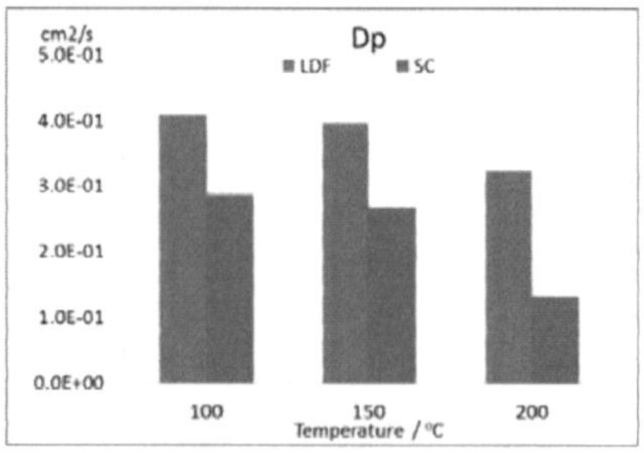

- ❑ AARD is 3.43%
- ❑ The Dp obtained with SC is smaller than that obtained with LDF.
- ❑ Non-isotherm models should be considered for better prediction

COLLEGE OF ENGINEERING AND COMPUTER SCIENCE | SYRACUSE UNIVERSITY

College of Engineering & Computer Science
SYRACUSE UNIVERSITY

MODELING: ISOTHERMS

GSTA Isotherm:

- ❑Multi-site model related to Heterogeneous Langmuir
- ❑Known to be effective for water vapor isotherms in MS3A[1]

$$q = \frac{q_{\max}}{m} \frac{\sum_{n=1}^{m} nK_n p^n}{1 + \sum_{n=1}^{m} K_n p^n}$$

GSTA Isotherm *(Llano-Restrepo and Mosquera, 2009)*
Code optimizes for $q_{\max}$, m, and each K_n simultaneously

GPAST model for co-adsorption:

- ❑ Provides the prediction of multi-component adsorption equilibria given the isotherm parameters and any temperature and pressure
- ❑ Can be realized with the MAGPIE (Multi-component Adsorption: Generalized Procedure for Isothermal Equilibria) application[2]

1.Lin, R., Liu, J., Nan, Y., DePaoli, D. W., & Tavlarides, L. L. (2014). I&ECR.
2.Ladshaw, A., Yiacoumi, S., & Tsouris, C. (2015). AIChE Journal.

CONCLUSIONS

❑Iodine adsorption on Ag°Z

- The optimal reduction conditions: 400°C, 24 hours.
- Optimal adsorption temperature is 150°C, where average I_2 loading is 13 wt% and Ag utilization based on chemisorption is 87%.
- SC model is capable to describe the uptake curves. The AARD is 1.47%.
- The presence of water in the gas stream reduces the iodine adsorption capacity.

❑Water adsorption on Ag°Z

- Uptake of water on Ag°Z is faster than iodine uptake under the conditions studied.
- The LDF and SC models are capable to predict the water uptake, AARD are 3.38% and 3.43%, respectively.
- Non-isotherm model should be considered for better prediction.

FUTURE WORK

❑Iodine/water co-adsorption experiments:

- T: 150°C; dew point: -40, -20, 10°C; Iodine conc.: ~50 ppmv
- The co-adsorption model will be verified.

❑Kinetics of I_2 reaction with Ag

- Experiments will be performed with Ag-coated glass beads
- Uptake curves at 100, 150 and 200°C will be obtained
- Kinetic model and parameters will be evaluated

❑Water adsorption on Ag°Z

- Non-isotherm kinetic models will be evaluated
- Equilibrium data will be analyzed with GSTA model
- I_2/H_2O co-adsorption model will be evaluated and verified

ACKNOWLEDGEMENTS

Funding support

- DOE NEUP; Standard Research Contracts: NE0008275

Collaborators

- Austin P. Ladshaw, Sotira Yiacoumi, and Costas Tsouris; Georgia Institute of Technology
- Jack D. Law; Idaho National Laboratory
- Jorge Gabitto; Prairie View A&M University

THANK YOU!

Pacific Northwest NATIONAL LABORATORY
Proudly Operated by Battelle Since 1965

Mixing Nuclear Waste Using Airlift Circulators: Performance Measurements and Parameter Optimization

AIChE Annual Meeting, Salt Lake City, UT
November 10, 2015

PP Schonewill, EJ Berglin, GK Boeringa, WC Buchmiller, CA Burns, and MJ Minette
Pacific Northwest National Laboratory

DH Alexander and LK Holton
Department of Energy, Office of River Protection

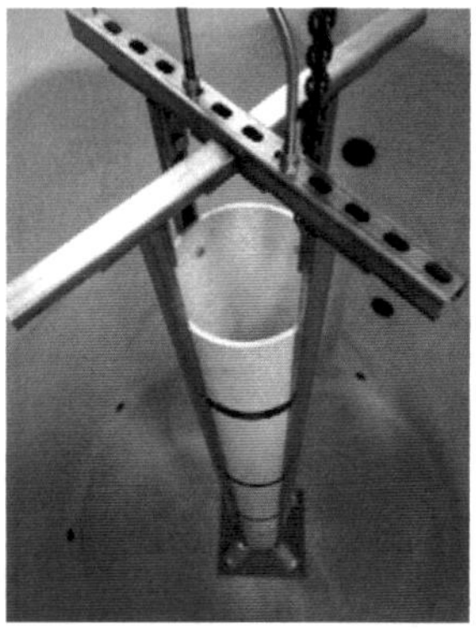

PNNL-SA-114265

Motivation for Study

Pacific Northwest NATIONAL LABORATORY
Proudly Operated by Battelle Since 1965

- Can an airlift circulator provide supplemental mixing to pair with a pulse-jet mixer (PJM) system?
  - Airlift circulator is a low maintenance alternative mixing system
  - Provide center upwell to promote mixing throughout the vessel height
  - Potentially reduce the amount of air required to promote mixing

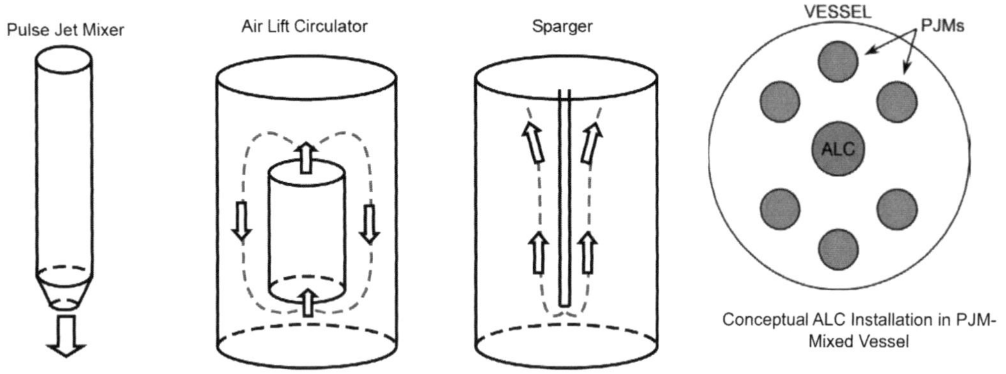

Conceptual ALC Installation in PJM-Mixed Vessel

2

Objectives of ALC Testing

Pacific Northwest
NATIONAL LABORATORY
Proudly Operated by Battelle Since 1965

- Demonstrate feasibility of technology for use in nuclear waste vessels
- Newtonian fluids
  - Lift/suspend settling particles in fluid
  - Distribute particles (homogeneously) in upper half of the vessel
- Non-Newtonian fluids
  - Yield and mobilize fluid, particularly in the upper half of the vessel
  - Minimize fluid "dead" zones

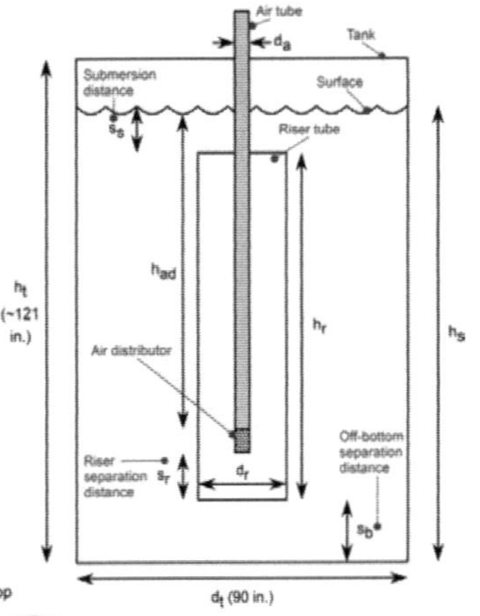

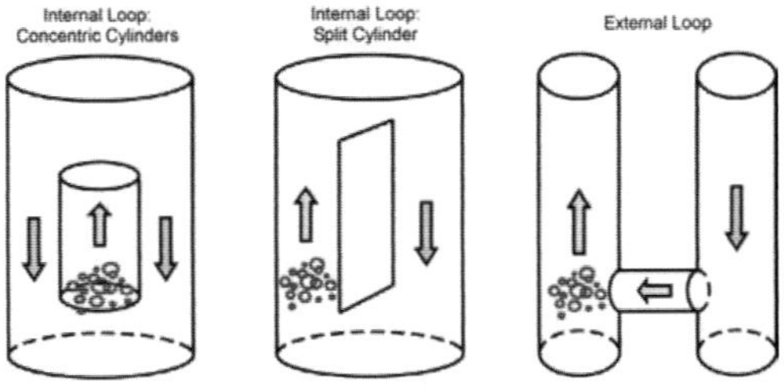

3

Velocity Performance Data

Pacific Northwest
NATIONAL LABORATORY
Proudly Operated by Battelle Since 1965

- Velocity data selected as a baseline performance measurement
- Collected using propeller instrument, located at riser tube outlet
- Averaged over 1 minute, then confirmed stable for additional minute.

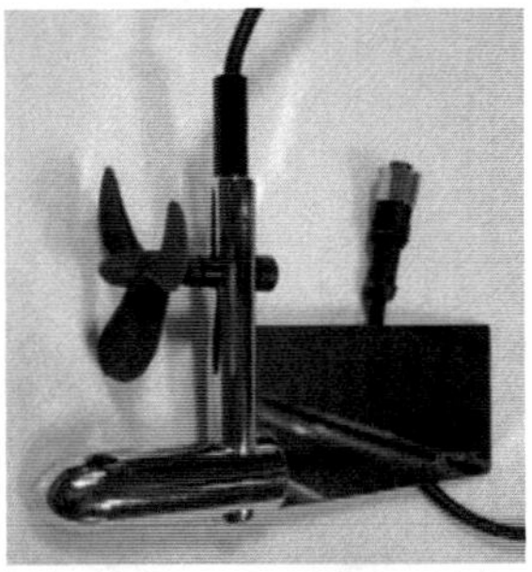

Air flow rate at standard conditions:

$$Q_{air,std} = Q_{air,rot}\left(\frac{P_{rot}T_{std}}{P_{std}T_{rot}}\right)^{1/2}$$

Estimation of liquid flow rate:

$$Q_{liquid} = \langle U \rangle \frac{\pi}{4}(d_r^2 - d_a^2) - aQ_{air,std}$$

4

Benchmark Testing

Pacific Northwest NATIONAL LABORATORY
Proudly Operated by Battelle Since 1965

Reproduce conditions of a test from a 1955 report (Cook and Waters, HW-39432).

1955 air distributor

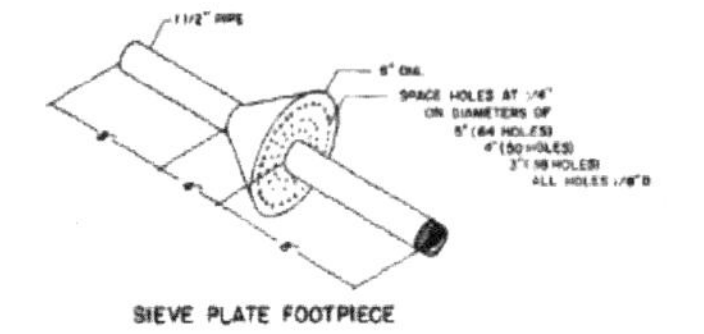

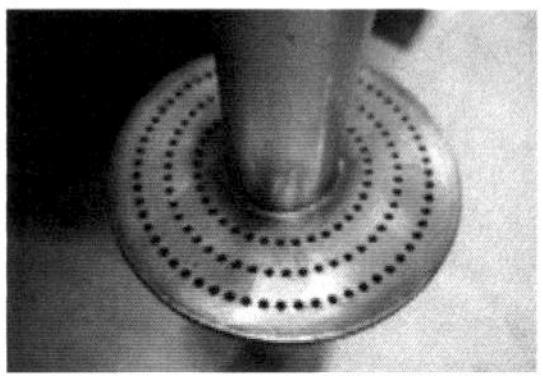

PNNL-fabricated replica

Compare estimates of liquid flow rate as function of air flow rate:

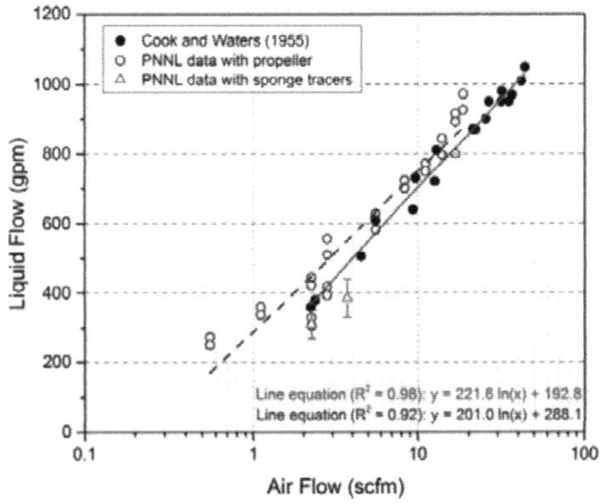

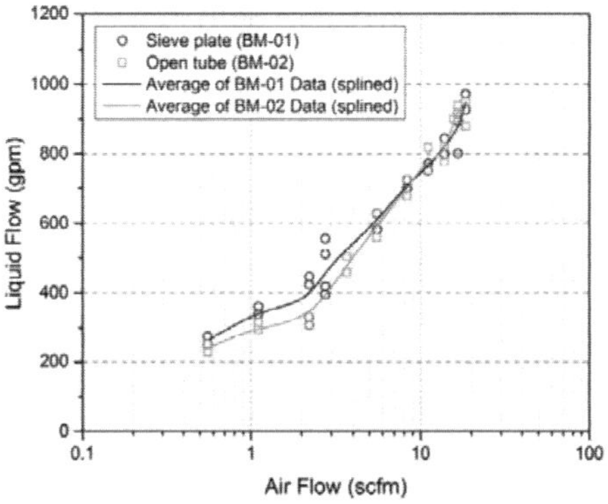

5

Velocity Data in Water

Pacific Northwest NATIONAL LABORATORY
Proudly Operated by Battelle Since 1965

Bello, RA, CW Robinson, and M Moo-Young. 1984. Canadian J. Chem. Eng., 62: 573-577.

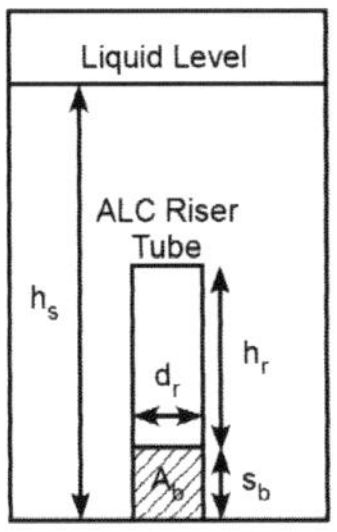

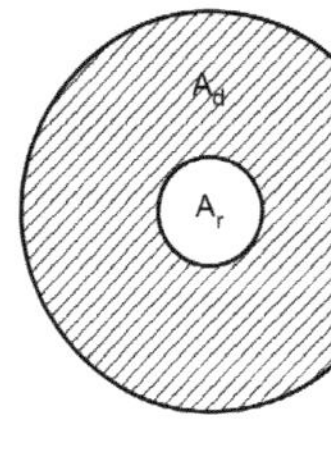

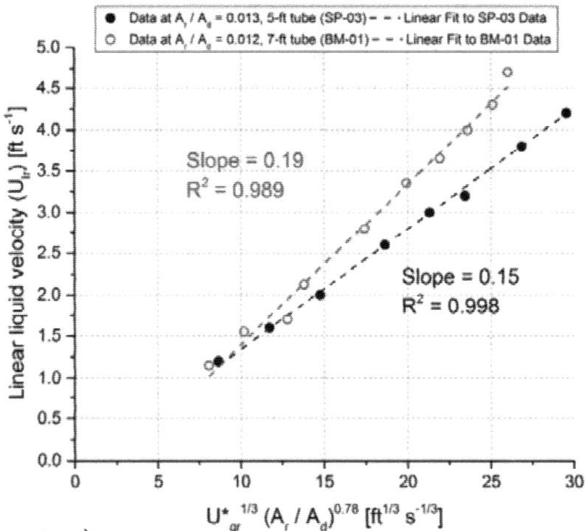

$$U_{lr} = \alpha \left(\frac{A_d}{A_r}\right)^{\beta} \left(U^*_{gr}\right)^{1/3}$$

$$U^*_{gr} = \frac{Q_{air,std} P_{atm}}{A_r (h_r - s_r) \rho_{sim} g} \ln\left(\frac{P_{atm} + \rho_{sim} g h_{ad}}{P_{atm} + \rho_{sim} g s_s}\right)$$

6

Summary Velocity Plot (Water)

Pacific Northwest NATIONAL LABORATORY
Proudly Operated by Battelle Since 1965

ID	d_r (in)	h_r (in)	S_b (in)
BM-01	9 7/8	83 15/16	13
SP-01	10 3/8	85 1/4	10
SP-02	10 3/8	85 1/4	5
SP-03	10 3/8	60 1/8	5 1/8
SP-04	10 3/8	36	5
SP-05	17.4	60	5
GP-02	3 15/16	56 5/8	2 7/8

$$U_{lr} = 1.71\left(U_{gr}^{*}\right)^{1/3}\left(\frac{h_r}{h_s - s_b}\right)^{0.7}$$

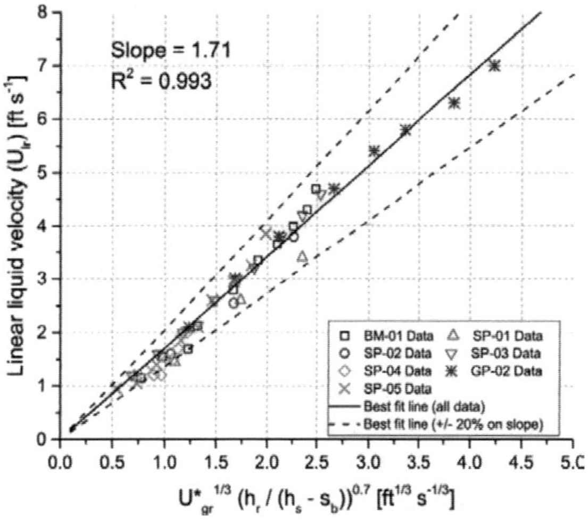

All A_r/A_d tested < 0.04

7

Velocity in Non-Newtonian Fluid

Pacific Northwest NATIONAL LABORATORY
Proudly Operated by Battelle Since 1965

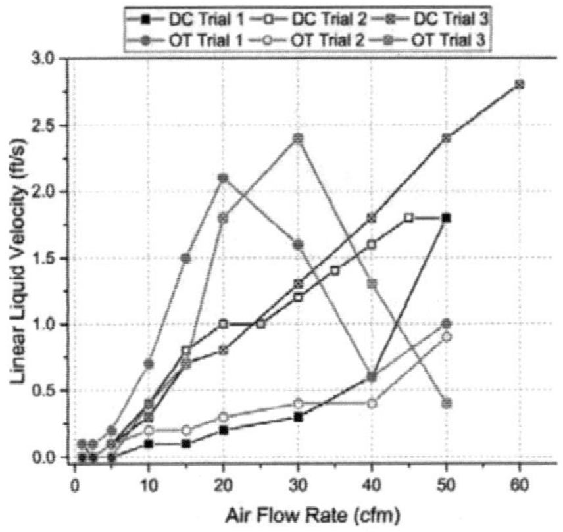

- Reproducing test results was a challenge
- Velocity instrument not sufficiently robust for non-Newtonian fluid

$$U_{lr} = \alpha\left(\frac{A_d}{A_r}\right)^{\beta}\left(U_{gr}^{*}\right)^{1/3}$$

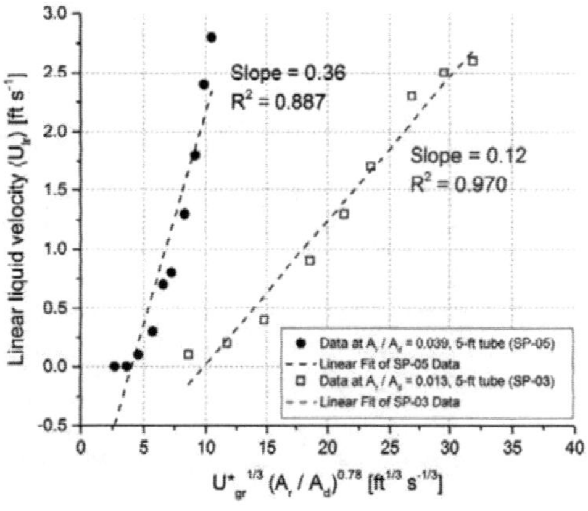

8

Newtonian Testing: Particle Lifting

Pacific Northwest NATIONAL LABORATORY
Proudly Operated by Battelle Since 1965

Objective: Isolate the ability of the ALC to transport and distribute particles from bottom of the tank to the top

Green particles:
500-750 μm glass spheres
density ~2.5 kg/L

Pea gravel: ~1/4 in (6 mm)
in maximum dimension
density ~2.7 kg/L

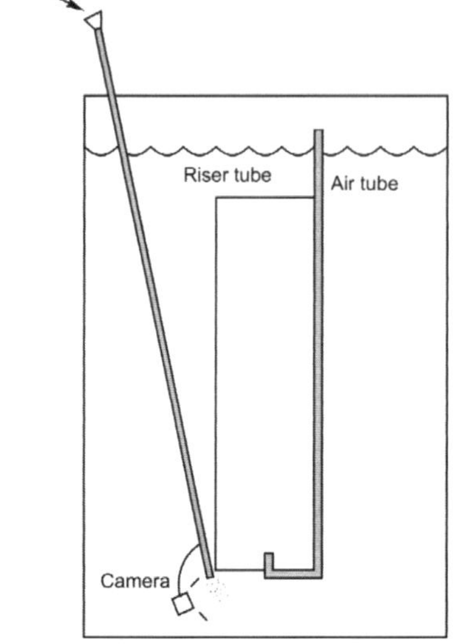

9

Video of Pea Gravel

Pacific Northwest NATIONAL LABORATORY
Proudly Operated by Battelle Since 1965

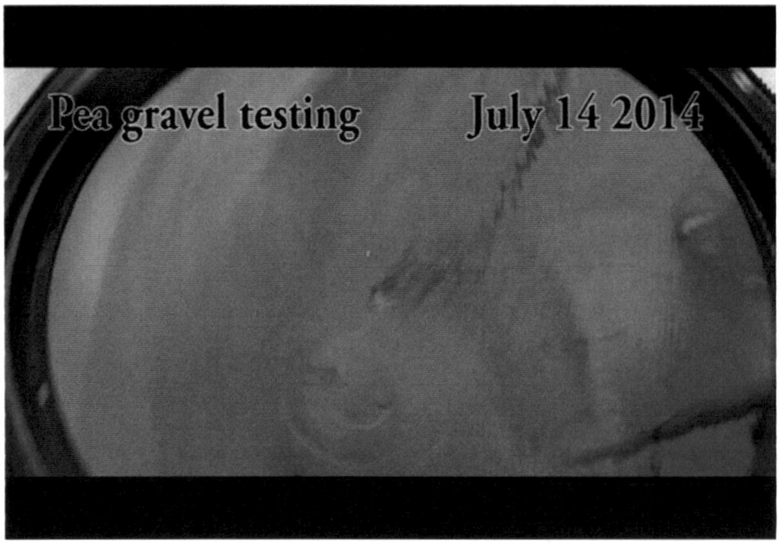

10

System Performance: Particle Capture

Pacific Northwest NATIONAL LABORATORY
Proudly Operated by Battelle Since 1965

Objective: Use a fixed particle capture location to quantitatively compare ALC configurations.

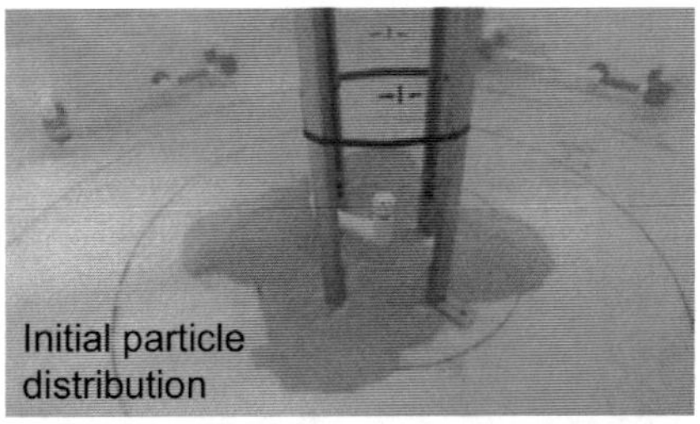

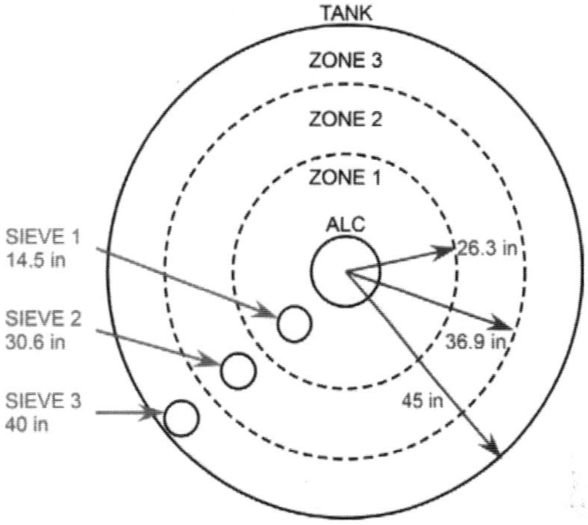

11

Particle Capture Test Results

Pacific Northwest NATIONAL LABORATORY
Proudly Operated by Battelle Since 1965

For ALCs tested at the same air flow:

- Larger diameter is better
- Pulsed air delivery is even better
- May be a optimum air flow rate depending on geometry

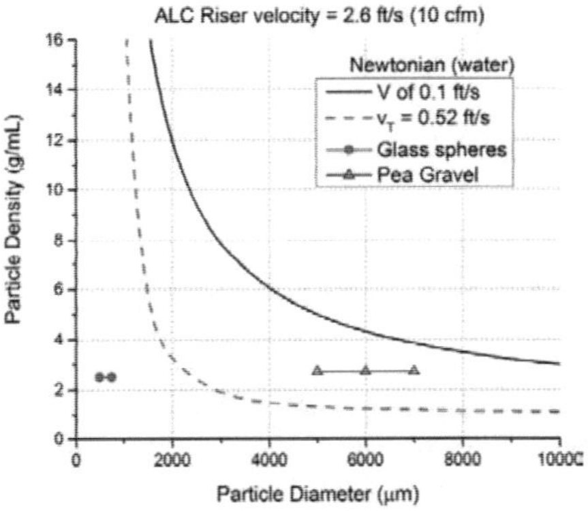

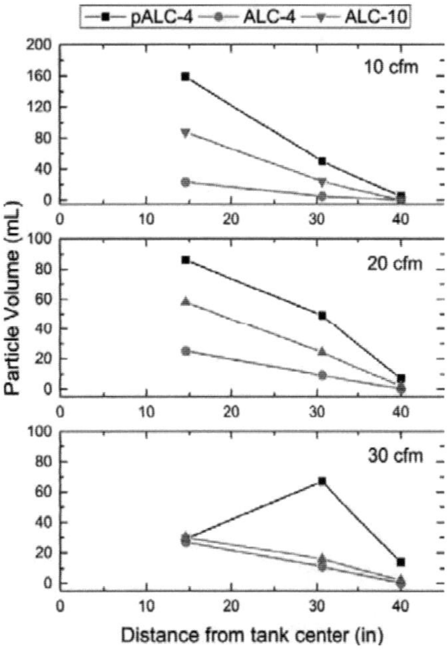

12

Non-Newtonian Testing

Pacific Northwest NATIONAL LABORATORY Proudly Operated by Battelle Since 1965

Non-Newtonian Fluid

- Use a clay slurry (80:20 kaolin:bentonite) with target Bingham parameters of 30 Pa (yield stress) / 30 cP (consistency).
- Actual clay properties:
  - Solids: 33.4 wt%
  - Density: 1.26 kg/L
  - Initial rheology: 22 Pa / 34 cP

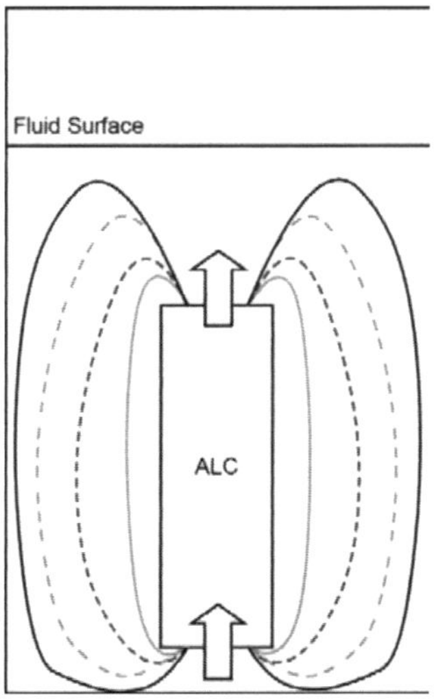

Does an ALC system fully mobilize non-Newtonian fluids, or are there significant dead zones?

13

Non-Newtonian Seed Particles

Pacific Northwest NATIONAL LABORATORY Proudly Operated by Battelle Since 1965

SEED PARTICLES

- Airsoft pellets (nominal properties):
  - 6 mm diameter
  - 0.12 g weight
  - 1.06 kg/L density
- Actual particles:
  - Density varied from 1.04 to 1.23 kg/L
  - One particle was 0.15 g and had a density of 1.40 kg/L

Yield-gravity number:

$$Y = \frac{\tau_0}{gl|\Delta\rho|}$$

Stability criterion for spheres is Y ~ 0.06.

For a bounding sphere in ALC testing, Y ~ 30. Y is larger for other spheres.

Spheres should remain immobile when fluid is at rest. Small positive buoyancy when fluid is yielded.

14

Non-Newtonian Test Method

Pacific Northwest NATIONAL LABORATORY
Proudly Operated by Battelle Since 1965

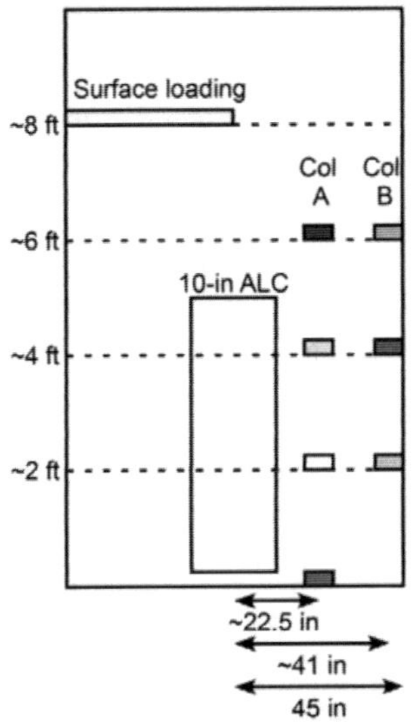

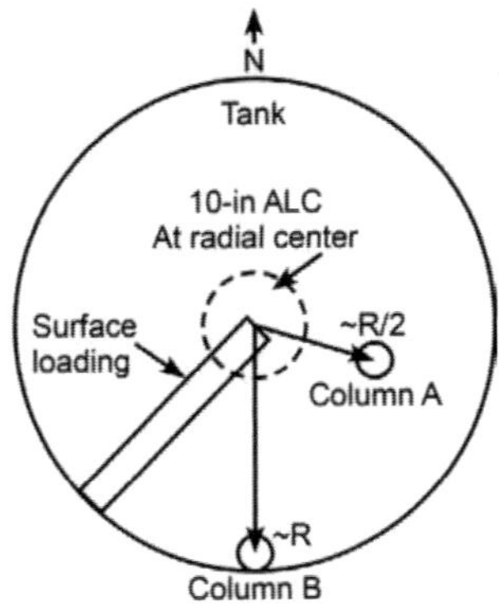

Preparing red particles for placement

Orange particles being placed

15

ALC Diameter Effect

Pacific Northwest NATIONAL LABORATORY
Proudly Operated by Battelle Since 1965

10-in diameter ALC

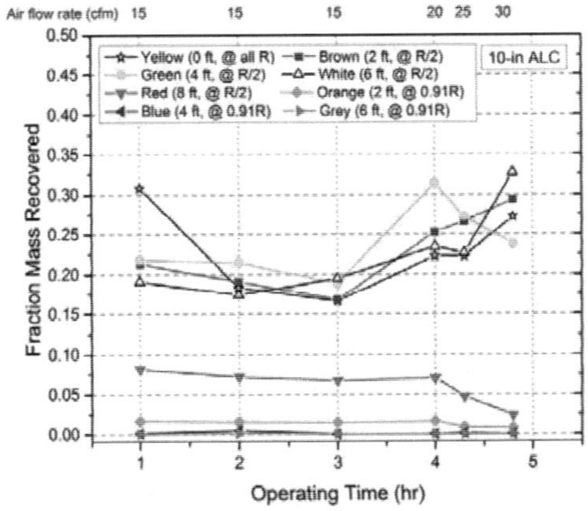

18-in diameter ALC

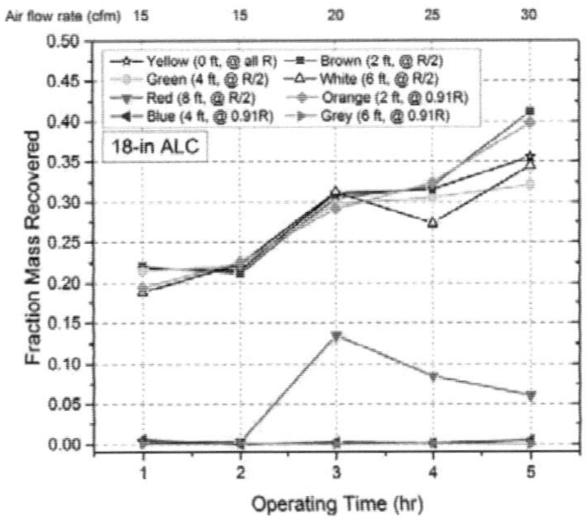

16

Qualitative Yielded "Cavern"

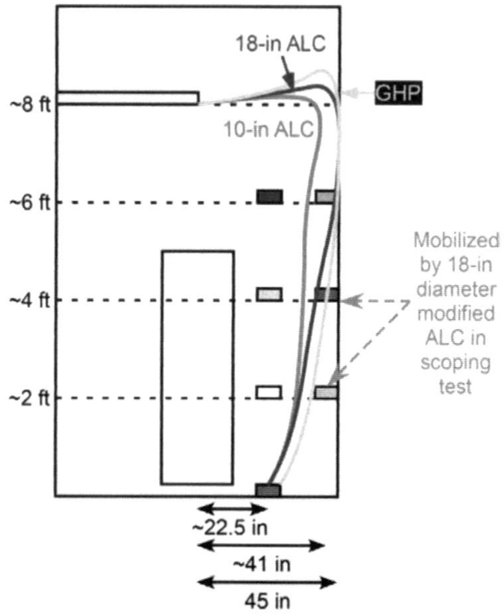

ALCs can yield non-Newtonian fluid out to the wall.

Variations in configuration can improve cavern size slightly.

Recovery of seeded particles suggests that the ALC establishes a recirculation pattern of yielded fluid in the vessel as expected.

Spargers also demonstrated to yield non-Newtonian fluid: do ALCs have any additional benefit?

17

Surface Zone of Influence

Not a significant difference between sparger and ALC in zone of influence.

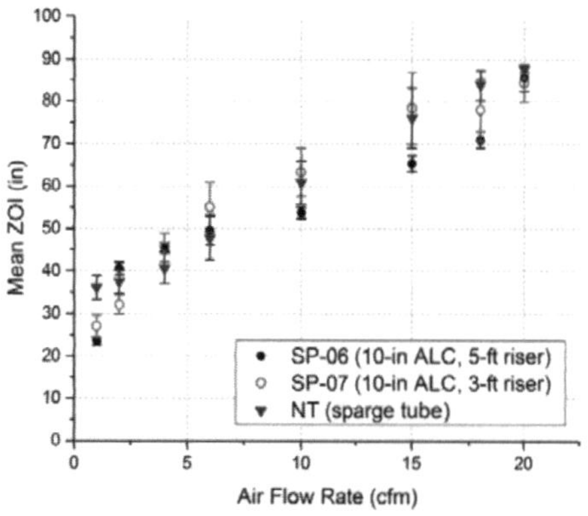

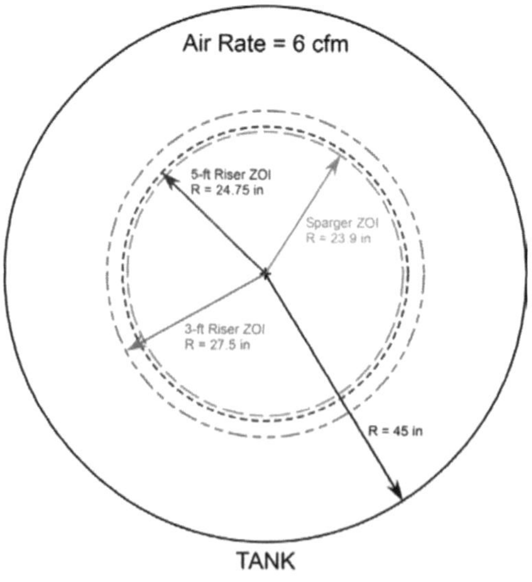

Determined using Ping-Pong ball tracers
Run device for 5 minutes
Measure from center to ball center

18

Sparger/ALC Comparison Test

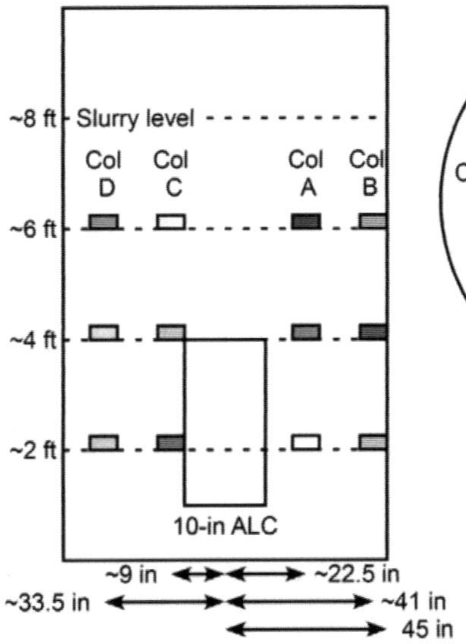

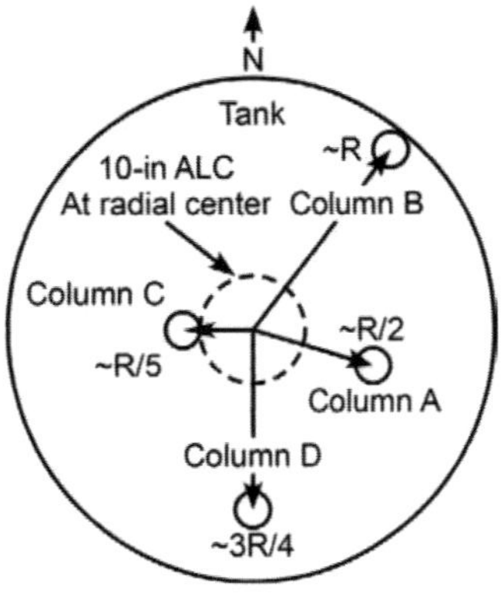

19

Mobilization Summary: 15 cfm

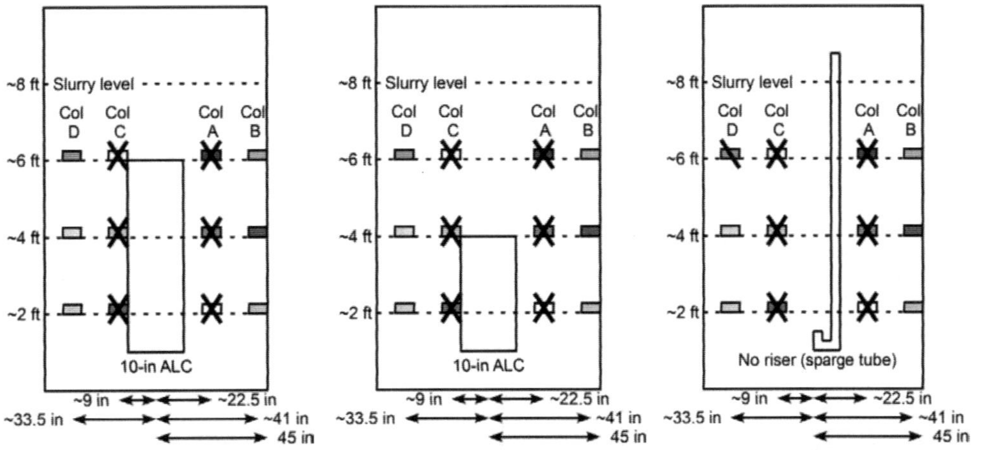

"X" indication: ≥ 10% of mass recovered when top foot of slurry sampled
"\" indication: ≥ 5% of mass recovered when top foot of slurry sampled

All configurations mobilized essentially the same particles; suggests the regions of mobilization are similar.

20

Mobilization Summary: 30 cfm

Pacific Northwest NATIONAL LABORATORY
Proudly Operated by Battelle Since 1965

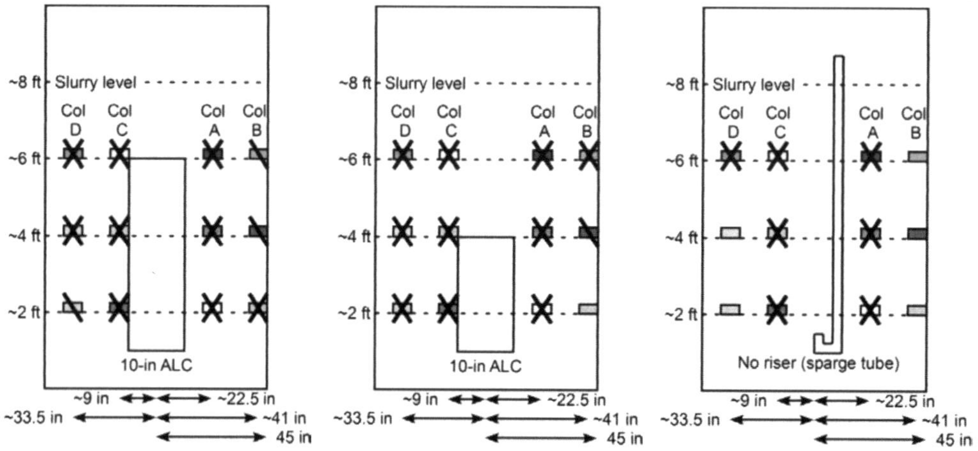

"X" indication: ≥ 10% of mass recovered when top foot of slurry sampled
"\" indication: ≥ 5% of mass recovered when top foot of slurry sampled

ALCs have larger region of mobilization; 3-ft riser marginally more than 5-ft

21

Multiple ALCs with Obstructions

Pacific Northwest NATIONAL LABORATORY
Proudly Operated by Battelle Since 1965

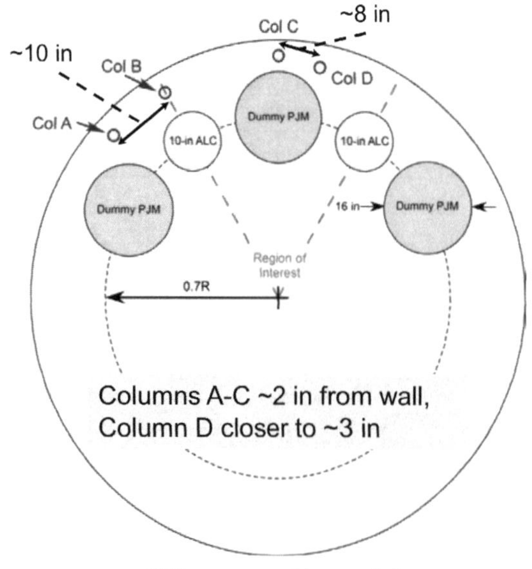

Diagram not to scale

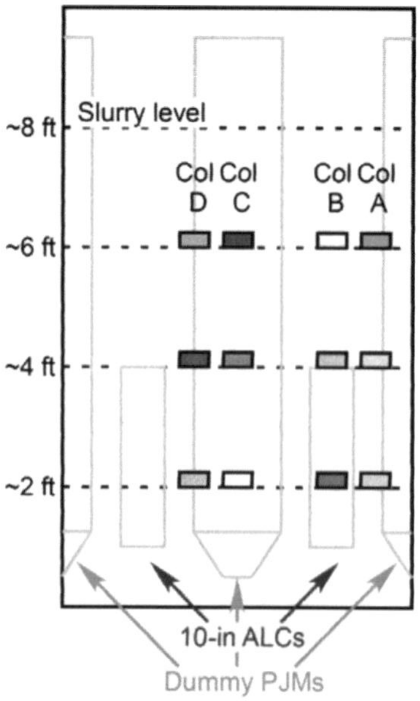

22

Evidence of Near-Wall Mobilization

Pacific Northwest NATIONAL LABORATORY
Proudly Operated by Battelle Since 1965

Near-wall regions are mobilized – but flow is slow!
Generally, 4-ft elevation is last to demonstrate mobilization.

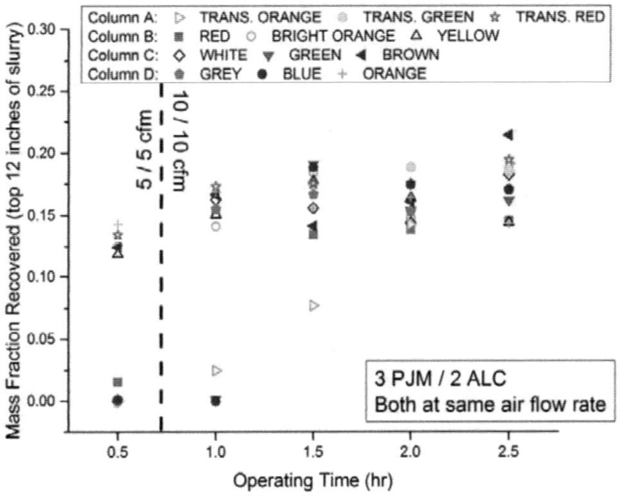

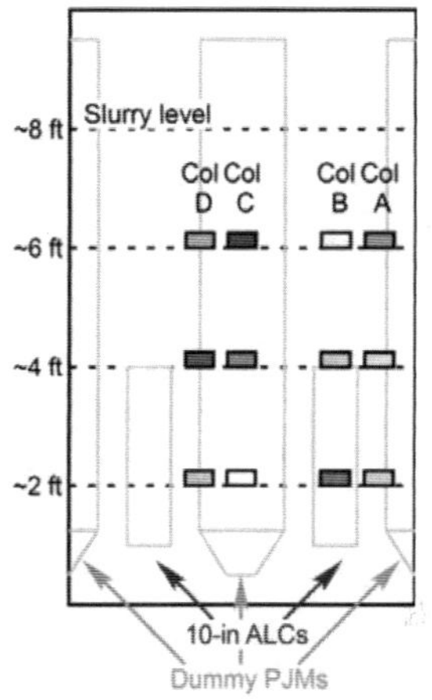

23

Air Lift Circulator (ALC) Summary

Pacific Northwest NATIONAL LABORATORY
Proudly Operated by Battelle Since 1965

- Empirical expression for liquid riser velocity in ALC developed
- Demonstrated particle lifting/suspension in Newtonian fluid:
  - Particle lifting threshold at ~5 cfm air flow
  - Limited mixing benefit at higher air flow rates
  - Capable of transporting solid particles to the vessel wall
- Circulation pattern creates a yielded cavern in non-Newtonian fluids
  - Cavern size grows with increasing air flow rate
  - Mobilization of near-wall/obstructed regions is achievable, but slow
- Application for mixing functions in nuclear waste vessels
  - Larger shear zone than spargers due circulating flow field
  - Could be modified for increased performance, i.e. use of a pulsed air flow in the riser tube
  - Other measurements needed to have data for scale-up/scale-down

24

Questions?

Pacific Northwest NATIONAL LABORATORY
Proudly Operated by Battelle Since 1965

25

Backup Slides

Pacific Northwest NATIONAL LABORATORY
Proudly Operated by Battelle Since 1965

26

Air Lift Circulator (ALC) Background

Pacific Northwest NATIONAL LABORATORY
Proudly Operated by Battelle Since 1965

- ALCs use air to drive/create circulation patterns in a vessel
- ALCs (or similar techniques) used in a variety of applications
  - Mining
  - Petrochemicals
  - Wastewater
  - Biological/biotechnology
- Successfully used to mix slurries with > 40 wt% solids
- Generally, large liquid circulation rates using small air flow rates

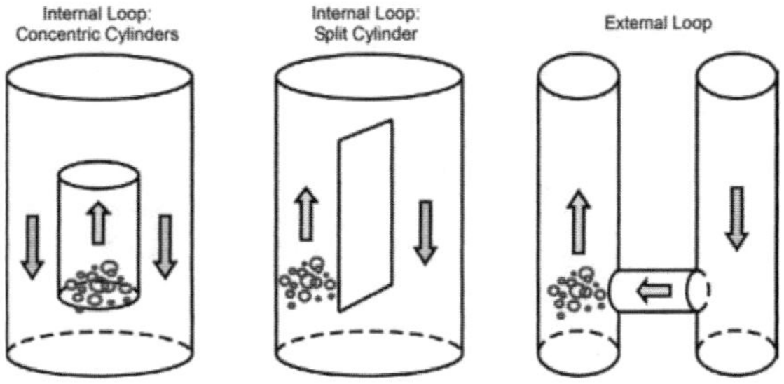

27

Additional Images ALC System

Pacific Northwest NATIONAL LABORATORY
Proudly Operated by Battelle Since 1965

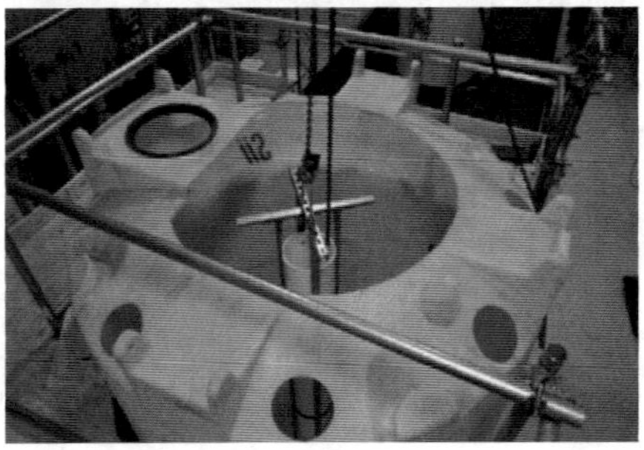

28

Additional Images (2)

Pacific Northwest
NATIONAL LABORATORY
Proudly Operated by Battelle Since 1965

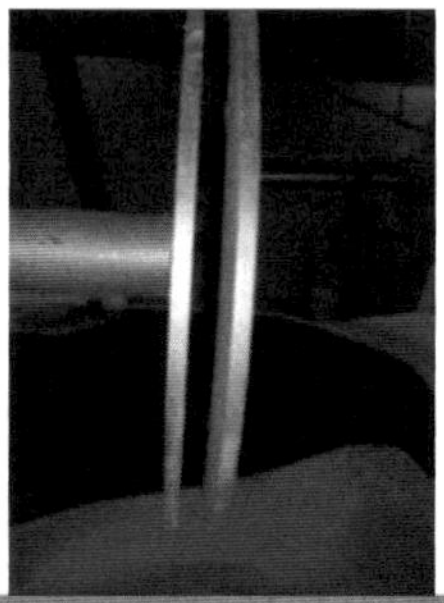

29

Particle Capture Test Results: Effect of Riser Tube Height

Pacific Northwest
NATIONAL LABORATORY
Proudly Operated by Battelle Since 1965

Riser tube length did not significantly effect performance

5-ft riser recommended based on expected higher fluid levels in future test vessels

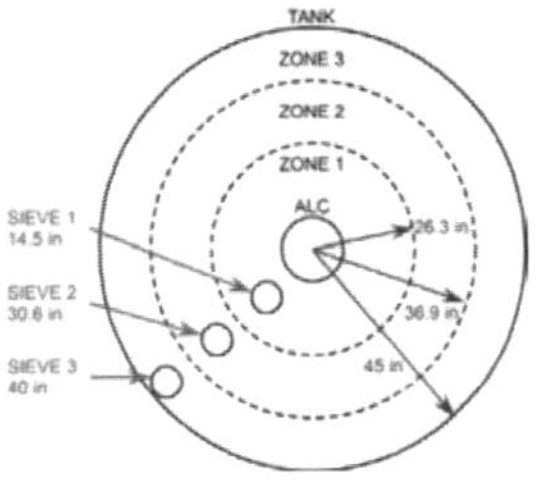

Test	Zone 1	Zone 2	Zone 3
SP-02 [7 ft riser; ~15 in below surface]			
SP-03 [5 ft riser; ~40 in below surface]			
SP-04 [3 ft; ~64 in below surface]			

30

Particle Capture Test Results: Effect of Air Flow Rate

Pacific Northwest
NATIONAL LABORATORY
Proudly Operated by Battelle Since 1965

Air flow rate did not significantly effect performance (above a minimum value)

Rates of 5 to 15 scfm recommended depending on fluid height in the vessel

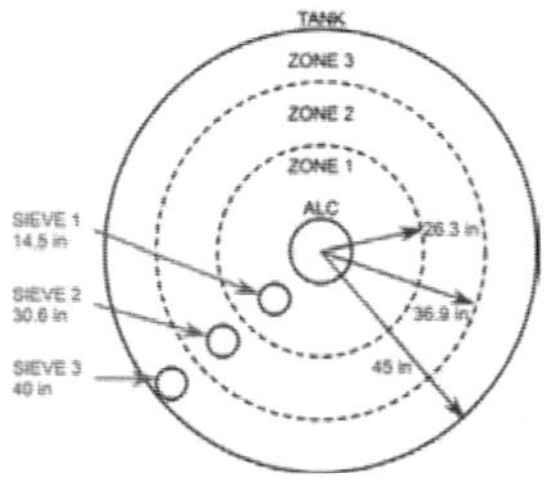

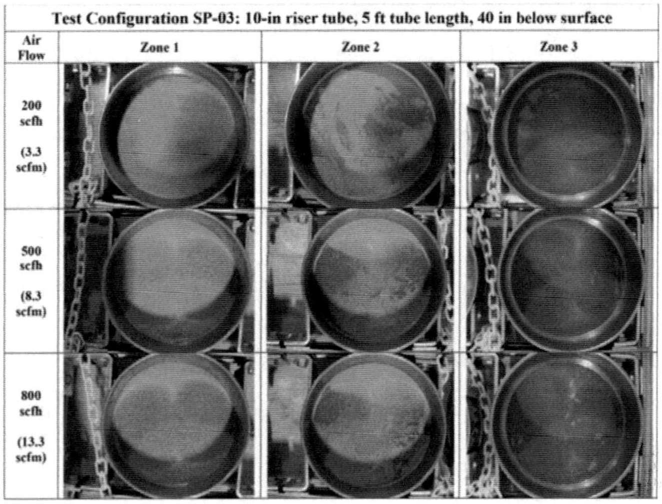

10-in and 18-in ALC Pump-Outs

Pacific Northwest
NATIONAL LABORATORY
Proudly Operated by Battelle Since 1965

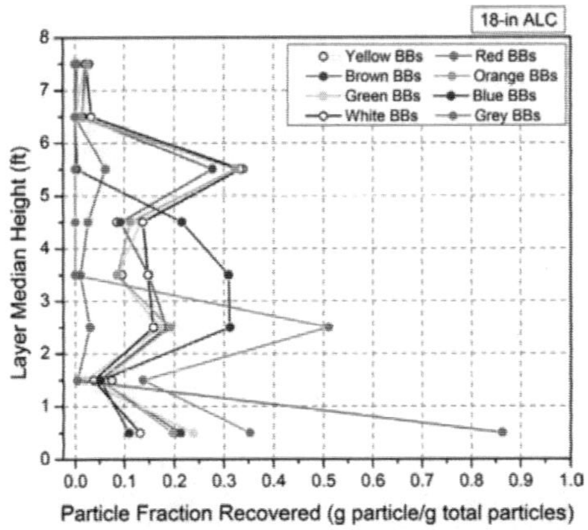

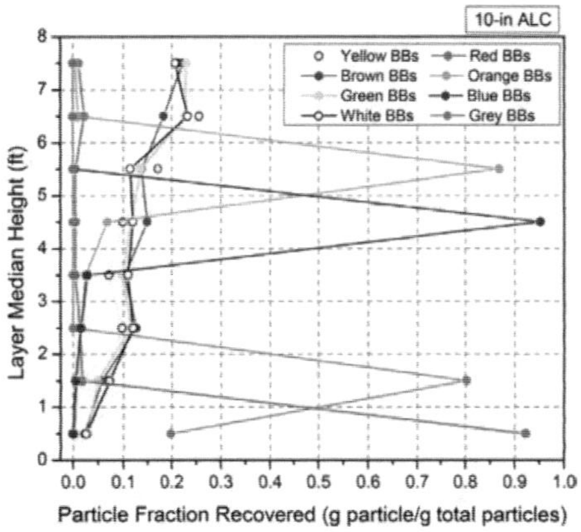

Key differences:

- Distribution of orange/blue/grey particles is less "focused" in 18-in ALC
- Spike in particles observed in layer just above outlet of 18-in ALC

3 PJM / 2 ALC Configuration

PJMs: non-functioning, hollow tubes of ~16 in OD; ~11 in conical section with 4 in nozzle

ALCs: 10 3/8 in ID; 3 ft length; metal tube

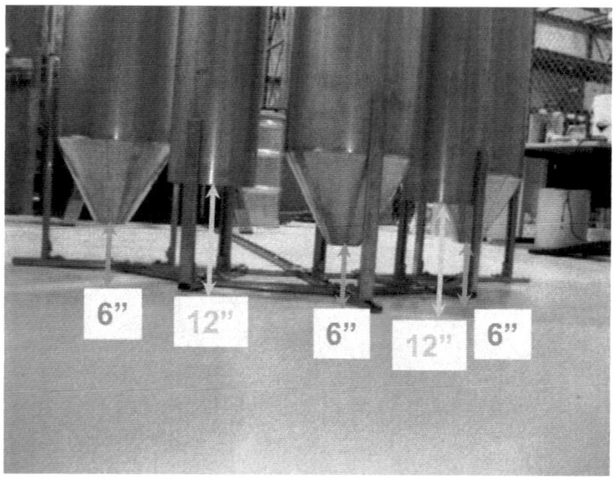

33

Particle Loading Technique (3 PJM/ 2 ALC)

Pacific Northwest NATIONAL LABORATORY
Proudly Operated by Battelle Since 1965

<u>Procedure</u>

1. ~100 g load of particles
2. Pour into tube, follow with a bit of slurry
3. Put particles in position with plunger
4. Fasten plunger to tube and remove

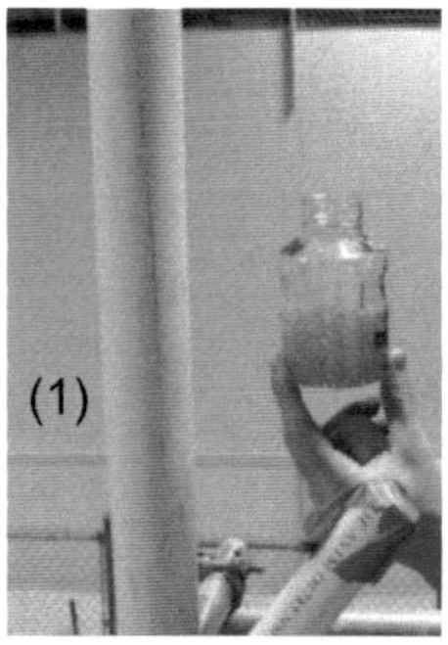

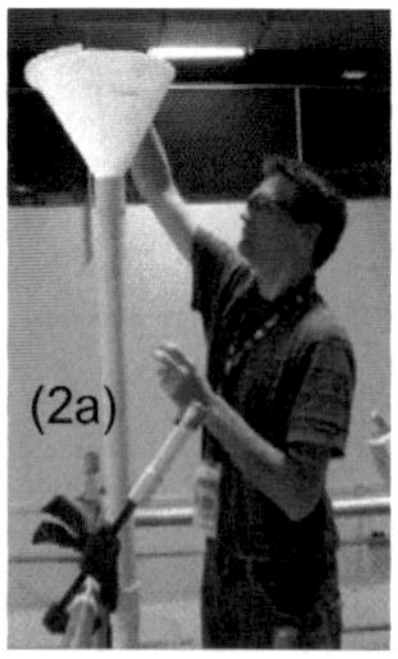

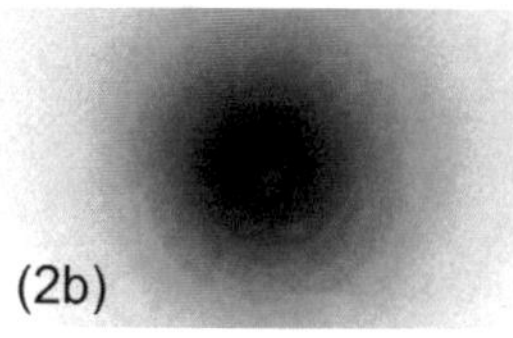

Result: spherical/cylindrical mass of particles 2-3 inches in dimension.

34

Collection of Velocity Data

Ultrasound Velocity Profiler (UVP) – Ultrasonic Doppler Velocimetry

- Manufacturer: MET-FLOW SA
- Model: UVP-DUO MX
- Serial No: 712
- Transducer Used: 2 Mhz

Installed on rigid pole, orientated vertically downward

Measures velocity in single direction projecting away/towards the transducer face – distance and resolution of measurement is determined by instrument parameter selection

35

Collection of Velocity Data, Continued

Transducer Depth: ~2-ft from bottom
Transducer Location: Behind center PJM
Orientation: Vertical, facing down
Data Collection: 2 min average
ALC Air Flow Rates: Both at 10 cfm

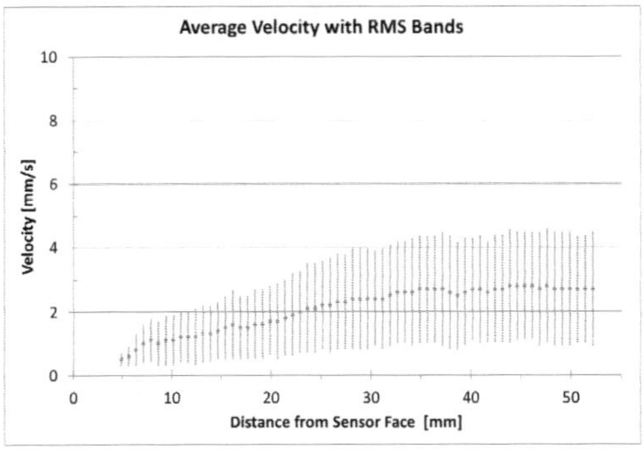

Transducer Depth: ~4-ft from bottom
Transducer Location: Behind center PJM
Orientation: Vertical, facing down
Data Collection: 2 min average
ALC Air Flow Rates: Both at 10 cfm

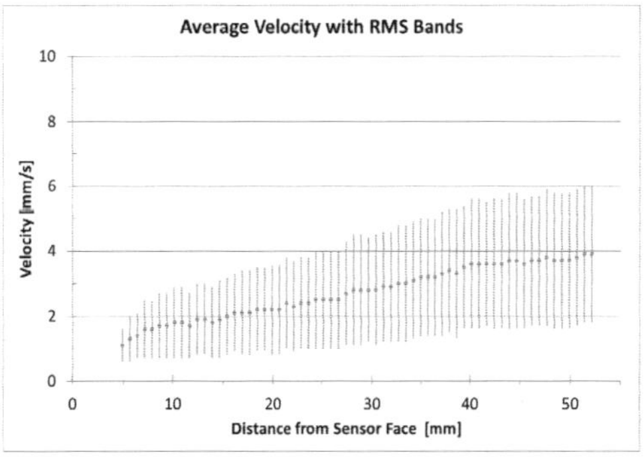

Velocities in confined regions (between PJM and wall) are small but detectable.

36

Newtonian Summary

Pacific Northwest NATIONAL LABORATORY
Proudly Operated by Battelle Since 1965

- Particle lifting and dispersion within ALC capability
  - Minimum air flow rate to consistently lift and transport particles is 5 cfm
  - ALCs of larger diameter/pulsed air flow are more effective for the same air flow rate
  - ALCs have small suction radii

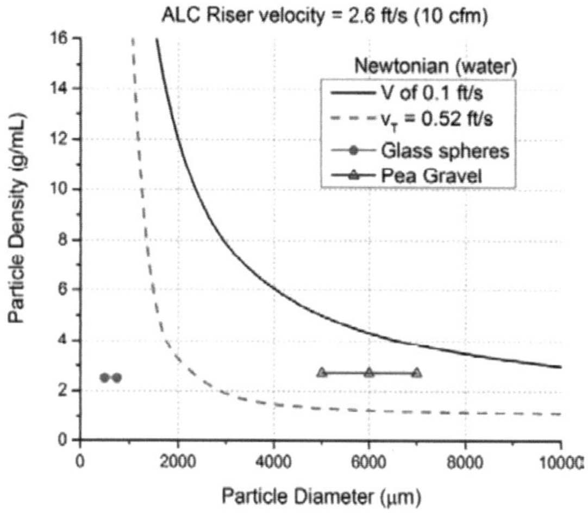

37

Experimental and Computational Investigations of Plenum-to-Plenum Heat Transfer and Gas Dynamics Under Natural Circulation in a Prismatic Very High Temperature Reactor

I. A. Said[1], M. M. Taha[1], S. Usman[2], and M. H. Al-Dahhan[1]

1) Department of Chemical Engineering, Missouri S&T, Rolla.
2) Department of Mining and Nuclear Engineering, Missouri S&T, Rolla.

Presenting Author: Mahmoud Taha Moharam (M. M. Taha)

MISSOURI S&T

Presentation Outline

1. Introduction
2. Motivation
3. Objectives
4. Work plan
5. Missouri S&T Scaled-down Facility (MSTF)
6. Proposed sophisticated techniques
7. Acknowledgement

Introduction

- The International forum in 2000 have released the Very High Temperature Reactors (VHTRs) is one of the most leading candidate for Next Generation Nuclear Plants (NGNPs).
- The design has high fuel efficiency for electricity production and because of the high gas coolant outlet temperature is optimal for industrial applications or commercial hydrogen production.

Introduction "cont."

- There are two versions of VHTRs:
  1. Pebble Bed Reactor (PBR)
  2. Prismatic Modular Reactor (PMR)
- The Gas-Turbine Modular Helium-cooled Reactor (GT-MHR), also termed Modular High Temperature Gas Reactor (MHTGR) is a candidate for the DOE-funded NGNPs prismatic modular reactors.

Introduction "cont."

- MHTGR (prismatic modular) design features:

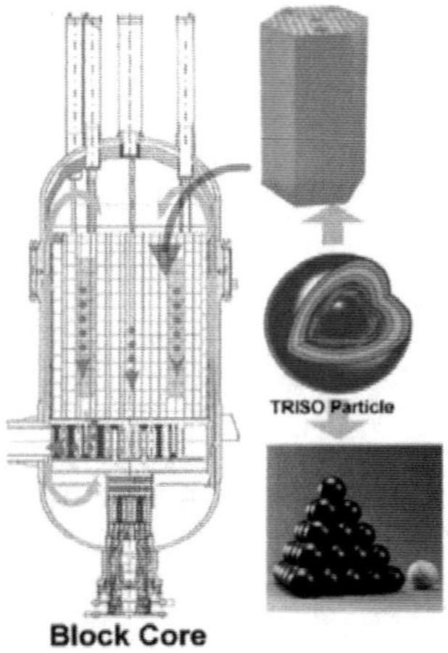

Figure 1. Features of Prismatic VHTR (yoon et al. 2011)

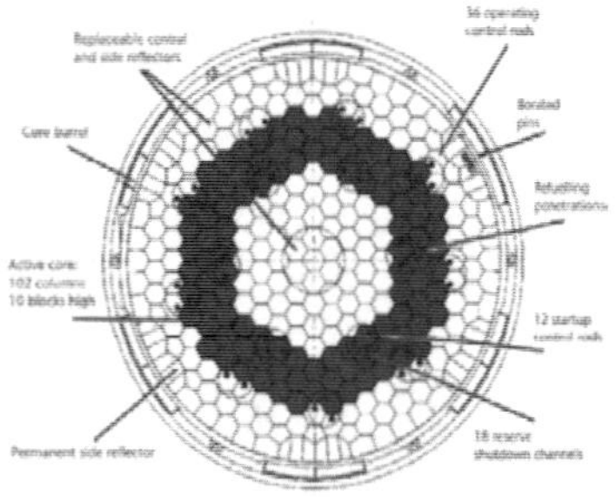

Figure 2. Core of prismatic CHTR (LaBar et al. 2004, Southworth et al. 2004)

5

Missouri University of Science and Technology

Motivation

- Computational Fluid Dynamics (CFD) has received a great deal of attention for understanding and predicting the flow behavior in VHTR, **BUT sufficient validation of these codes has not been carried out.** (Lee et al. 2014, McVay et al. 2015)
- Lack in a detailed experimental study in the open literature for experimental work that addresses the gas dynamics and heat transfer in the prismatic core during natural circulation by using advanced novel techniques. (DOE 2016 Solicitation)

6

Objectives

- Design and develop a separate-effects experimental setup mimicking prismatic modular reactor.
- Advance and address experimentally and computationally the gaps in scientific and engineering knowledge of natural circualtion.

"benchmark data is needed to validate CFD simulations while there were no experimental test facilities to provide this"

McVay et al. 2015

Objectives

- Provide benchmark data for validation of CFD commercial codes.

"benchmark data is needed to validate CFD simulations while there were no experimental test facilities to provide this"

McVay et al. 2015

Work plan

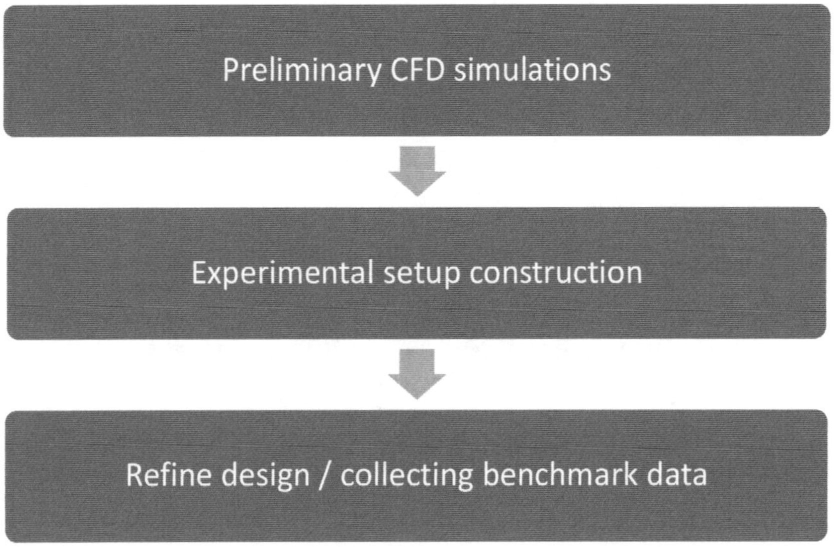

9

Missouri University of Science and Technology

Missouri S&T Facility (MSTF)

- The same scaling criteria used for building the High Temperature Test Facility (HTTF) at Oregon State University (OSU) with reference to MHTGR are employed.
- These scaling criteria have shown a successful similarity between HTTF and MHTGR. (Brian M. King 2012, R.R. Schultz et al. 2010)

10

MSTF "cont."

- MS&T dual-channel Facility *"Phase-I"* is a 1:4 scaled down facility based on OSU-HTTF design as shown below.

Table 1. Dimensions of MSTF

Parameter	Missouri S&T Facility
Tube Diameter (inch)	0.625 (same as HTTF & MHTGR)
Tube length(inch)	23.228 (1/4 of HTTF channel length)
Core diameter(inch)	11.78 (1/4 of HTTF core diameter)
Upper plenum height (inch)	9.41 ((1/4 of HTTF plenum height)
Outer vessel diameter (inch)	15 (1/4 of HTTF vessel diameter)
Number of channels	Two (one for up-ward and one for down-ward flow)

MSTF "cont."

Figure 3. Physical picture of MSTF experimental setup

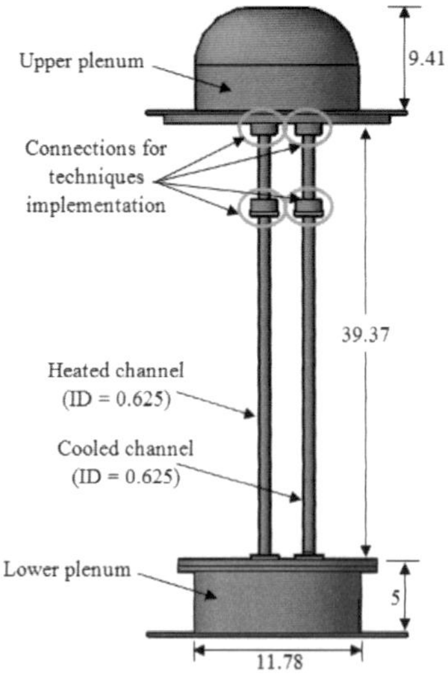

Figure 4. Schematic diagram of MSTF

Sophisticated Measurement Techniques

- There are many invasive and noninvasive techniques available in the Dr. Al-Dahhan labs such as:

1. Radio-isotopes noninvasive techniques
   - Gamma Ray Computed Tomography (CT)
   - Radioactive Particle Tracking (RPT)
   - Gamma Ray Densitometry (GRD)

Sophisticated Measurement Techniques

2. Non-Radioisotopes based invasive and noninvasive techniques
   - 1-point, 2-point and 4-point optical probes
   - Advanced dynamic gas and liquid tracer
   - Heat transfer probes
   - Hot wire anemometry (Constant Temperature Anemometer - CTA)

Sophisticated Measurement Techniques

- In the current study the following techniques are planned to be used:
  1. Heat transfer sensors.
  2. Hot wire anemometry
  3. Dynamic gaseous tracer

MISSOURI UNIVERSITY OF SCIENCE AND TECHNOLOGY

Heat Transfer sensors

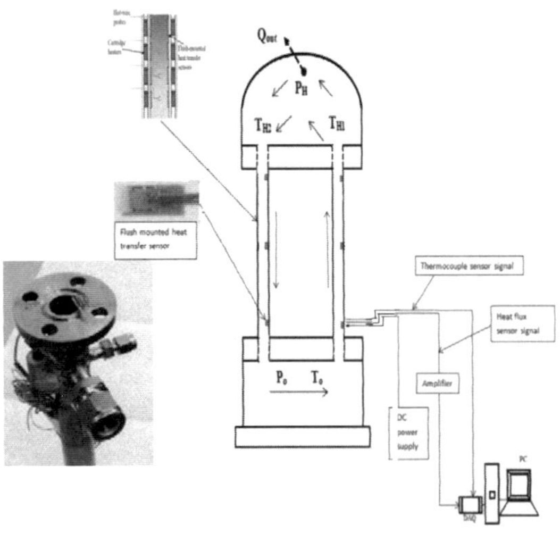

Figure: Schematic diagram of proposed location of heat transfer probes

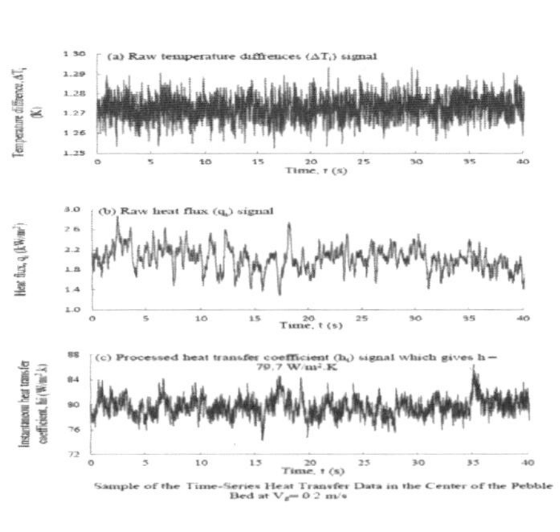

Figure: Sample of time-series heat transfer data

Hot Wire Anemometry

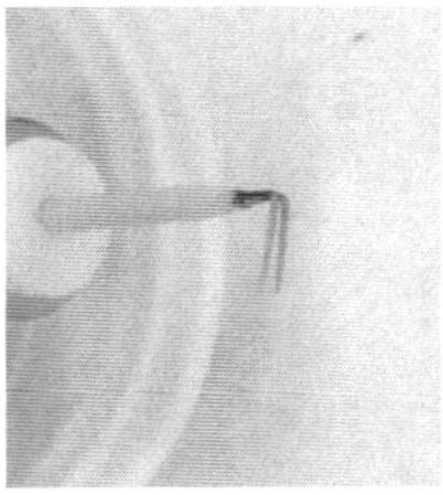

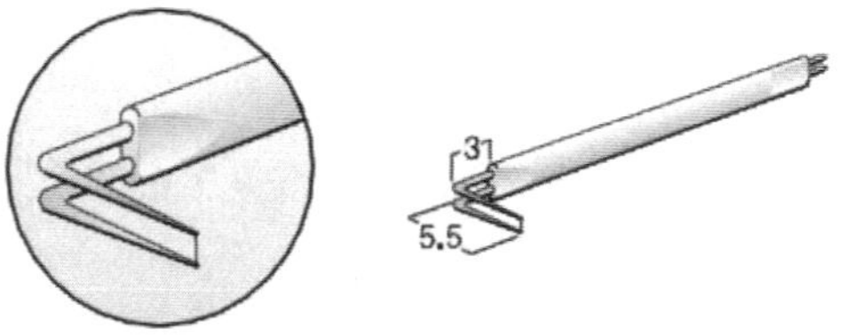

55P14

90° sensor perpendicular to probe axis

Figure: CTA used in the current study

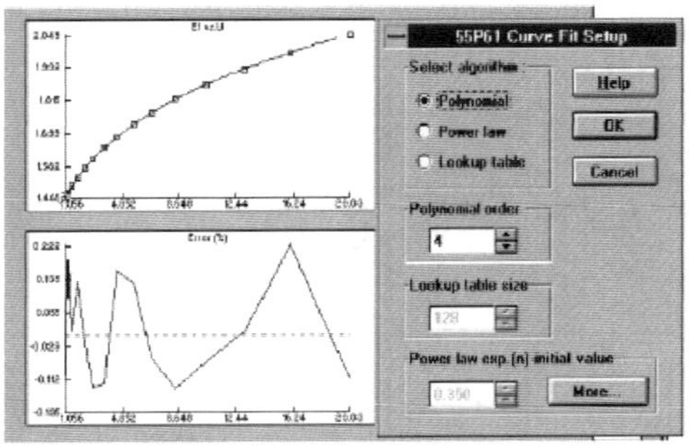

Figure: Calibration results

Dynamic Gas Tracer

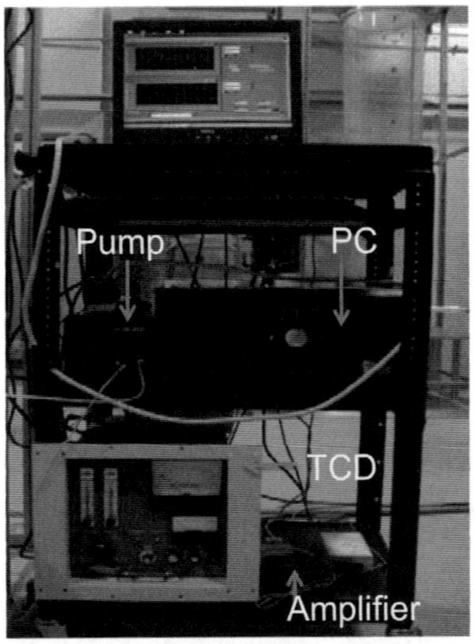

Figure: Gaseous tracer data acquisition system and data analysis

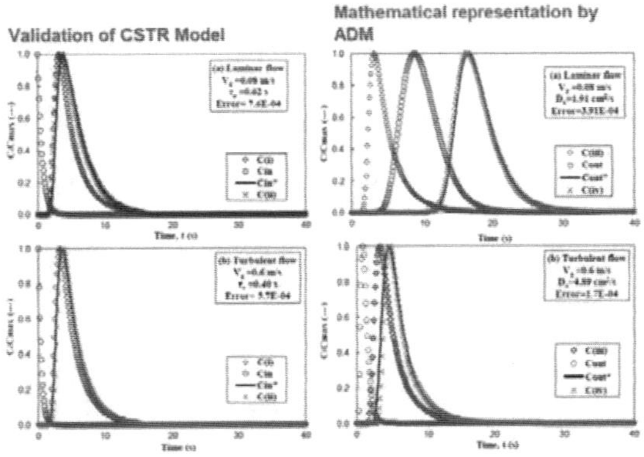

Figure: Sample of gas tracer results

Outcomes/Impact

- Developing flexible, separate-effects experiments to investigate intra-core natural circulation which complement the experiments of HTTF.
- Developing an integrated sophisticated techniques that are implemented for the first time in such a study.
- Obtaining detailed benchmark data for CFD codes validation.
- Advance the knowledge and understanding of P2P natural circulation thermal hydraulic phenomena.

Missouri University of Science and Technology

Acknowledgment

The authors acknowledge the financial support provided by the U.S. Department of Energy project number (NEUP 13-4953 (DE-NE0000744)) for the 4th generation nuclear energy, which made this work possible.

THANK YOU …

Team members	Emails
I.A. Said	iasvx4@mst.edu
M.M. Taha	mmt8t6@mst.edu
S. Usman	usmans@mst.edu
M.H. Al-Dahhan ***(Project PI)***	aldahhanm@mst.edu

What ever happened to Thorium?

Or why not U-232?

U what you might ask!

- We have U-233, U-235, U-238, etc.
- So why not go from Th-232 to U-233?
- At one time Th to Uranium was considered to be a very viable process.
- But then again a whole lot of promising energy projects went down the tubes after 3-Mile Island.
- And Jimmy Carter our Nuclear Engineer President!

Is Thorium too nasty for REE Recovery?

- Rare Earth Elements or REE's are highly valued minerals.
- The question is will Thorium, which is found with REE's make their recovery process too difficult?
- As you know Thorium is radio active with a very, very long half life!
- Should this radio activity be a deterent to reclaiming valuble REE's?

What about U 232?

- Since Jimmy Carter killed the U.S. Nuclear Industry, Various forms of potential power sources have gone into semi-oblivion!
- At one time Thorium to U 232 was an exciting prospect.
- Very little is heard about U 232.
- When I was at PORTS in Piketon, OH, in the late 90's, we were the only DOE plant refining U-235.

PORTS

- Oak Ridge was shut down and now is PORTS. All that is left is vast clean-up of these plants, along with SRS and the Washington State plants.

- Clean-up was the sole task of our engineering group at PORTS.
- Our experimental centrifuges were barely run, if ever!

Thorium to U-232

- Because of all the need for REE's, Thorium and its inclusion in most REE deposites, is a natural byproduct of REE mining.
- While China has vast REE reserves, the USA also has REE reserves.
- One of the prime area's is in South East Missouri, namely Sullivan County.

Pea Ridge Mine

- Pea Ridge was a Magnetite Iron Ore mine operating in Sullivan, MO.
- It is now closed, and flooded.
- But thousand of metric tons of tailing are available for processing.
- These tailings contain all REE's plus Thorium!

Pea Ridge

- For some reason Geological surveys seem to ignore Pea Ridge in favor of area's such as Mountain Pass in California!
- According to the USGS of 2014, the only source of REE's is CHINA!!
- Is this anyway to secure our Nations future?

Uses for REE's

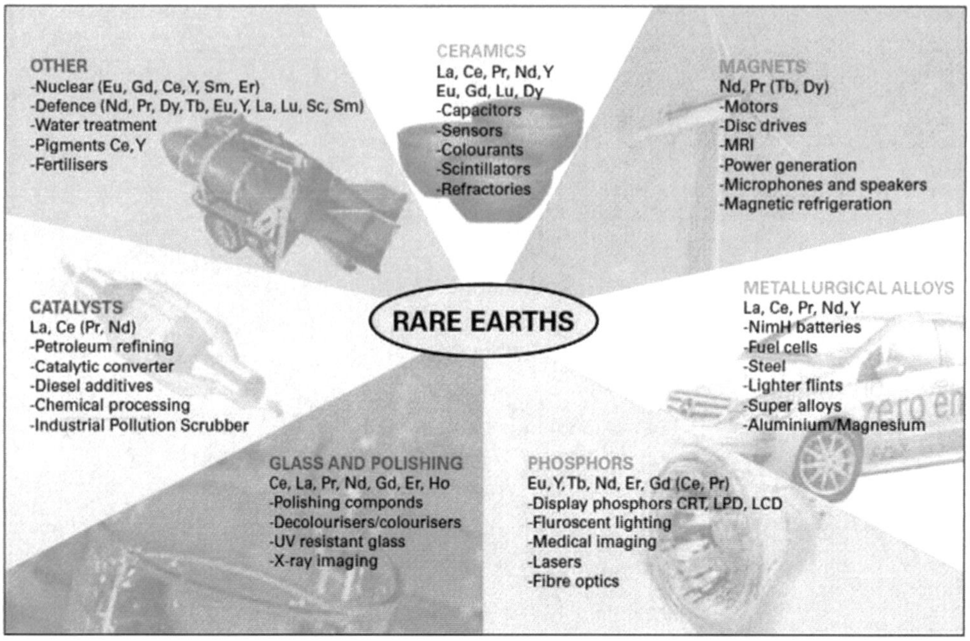

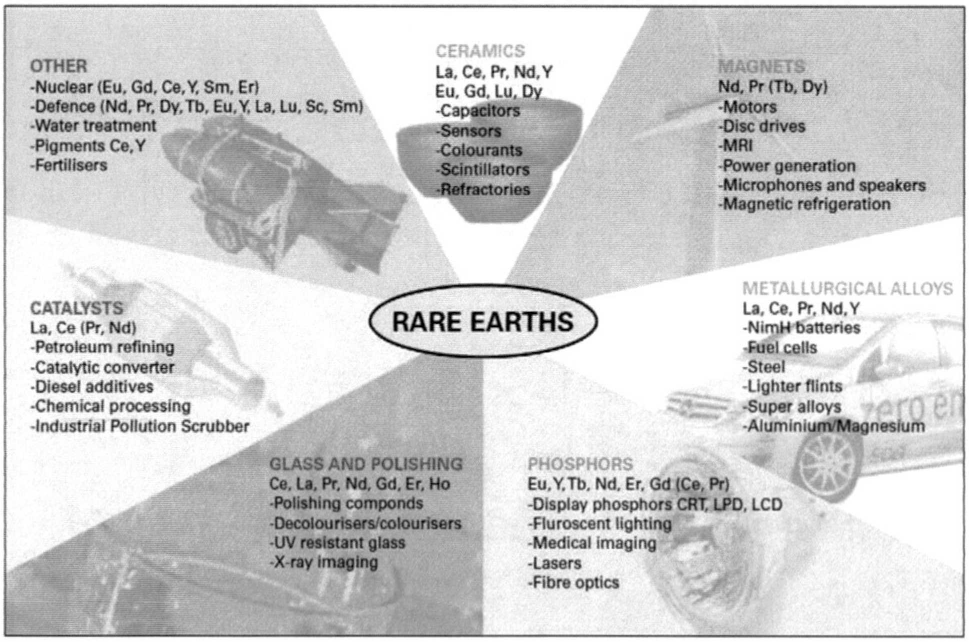

Periodic Table

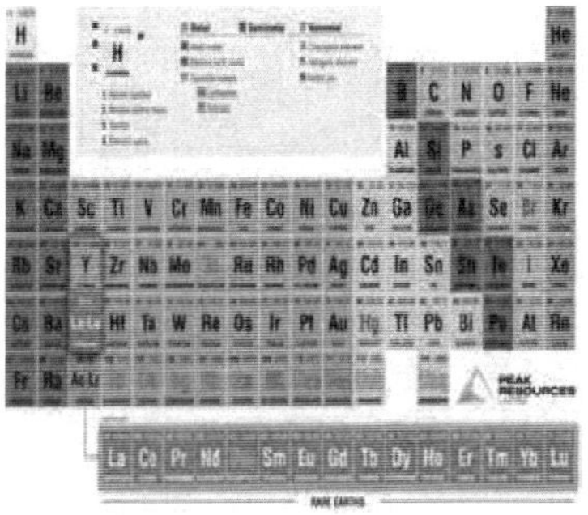

REE Production (1800-1991)

- **Figure 1.** Global rare earth element production (1 kt=106 kg) from 1950 through 2000, in four categories: United States, almost entirely from Mountain Pass, California; China, from several deposits; all other countries combined, largely from monazite-bearing placers; and global total. Four periods of production are evident: the monazite-placer era, starting in the late 1800s and ending abruptly in 1964; the Mountain Pass era, starting in 1965 and ending about 1984; a transitional period from about 1984 to 1991; and the Chinese era, beginning about 1991.

Periodic Table

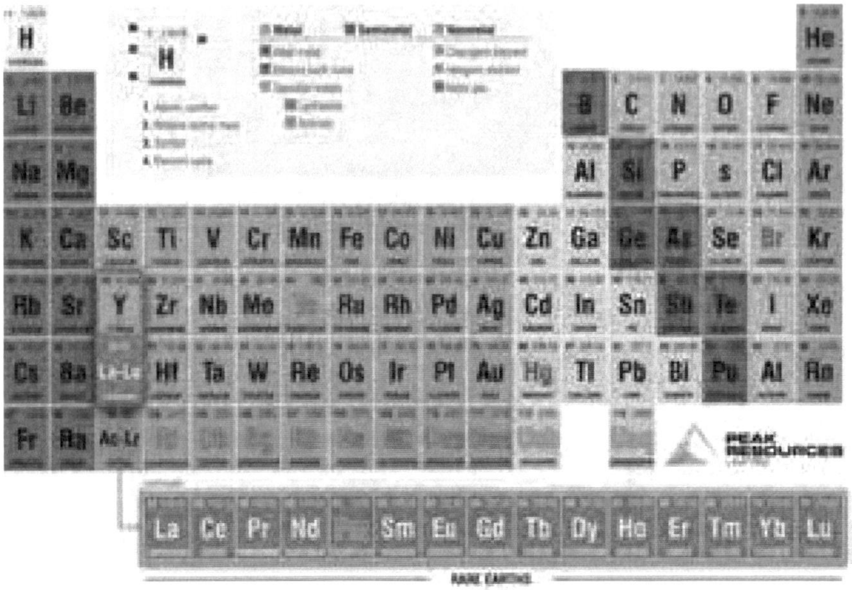

REE's Mt. Pass,CA;China;Bokan,AK; P.R,MO

Mtrc.tons	Mt. Pass	China	Bokan	Pea R
La	743K	164K	1.7K	351K
• Ce	1,420K	253K	5.3K	579K
• Nd	247K	116K	2.3K	105K
• Ga	4.7K	25.7K	0.7K	11.8K
• Y	2.6K	147K	3.7K	7.3K

- * Note: Pea Ridge tonnage is split about 50/50 between Breccia Pipes and Apatite tailings for most elements listed!

Why should Red China control the World's supply of REE's???

- Because they are our best buddy's!
- We can always trust people who want to take over the South China Sea!
- Because they could bankrupt the U.S.A. anytime they want!
- Because they hack our government and major corporations anytime they feel like it!
- Because they have veto power over progress made at the U.N.!

World Production of REE's Scale is just an approximation of relative outputs!

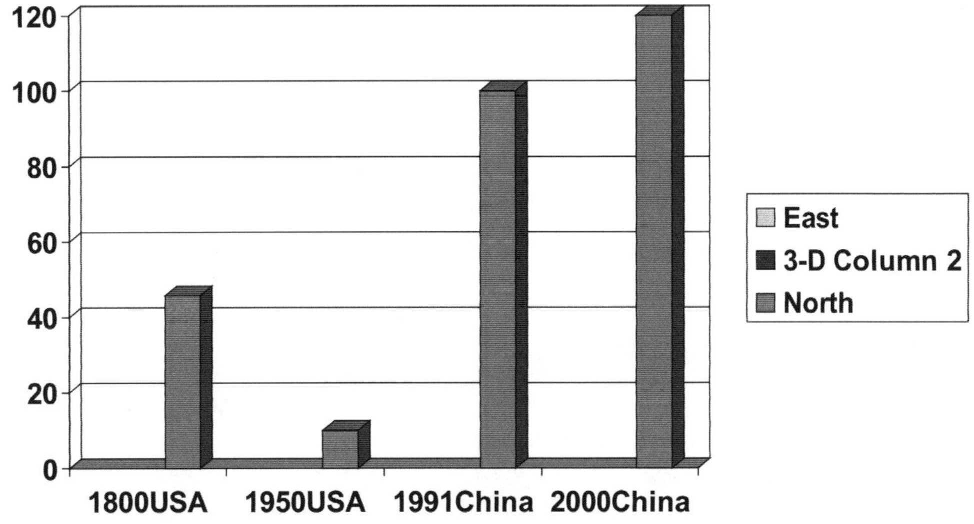

Comments from attendees!

- Yes, the fellow in the third row!
- OK so there's no one in the third row!
- Is Dr. Sheldon Cooper in the house?
- Anyone, PLEASE!! Jim help me out here!

Mr. Pea Ridge-- Jim Kennedy!

Kennedy

Processing of Thorium

- First let me define two terms:
- 1. Apatite- Mine tailings from Pea Ridge Iron ore (magnetite) mine.
- 2. Breccia – as in Breccia pipe – a tubular volcanic eruption of various diameters and in Pea Ridge's case several hundred feet long; that contain Substantial Concentrations of REE's plus Th, Au & Ag!

Processing of Thorium

- Pea Ridge contains both Apatite and Breccia.

- Having visited Pea Ridge in 1993 my primary interest, as was the companies, was The Breccia Pipes.
- Since production halted in about 2,000; the mine has flooded, making Apatite the much more appealing source of REE's and Thorium!
- The Apatite tonnage is about 500 to 600K metric tons.

408881 Coproduction of Chemicals: Improving the Value Proposition of Nuclear Power

Abstract:

Use of the heat from nuclear reactions has been proven over decades of commercial operation to be a reliable large-scale carbon-free source of electricity. Modern reactor designs and advanced fuel cycles are resistant to weapons proliferation and contain inherently safe technologies with far less waste than conventional reactors. Despite a consistent safety record and the lack of carbon-free base load alternatives, the nuclear power industry in western democracies has been unable to remain competitive with fossil fuel-based power production. There is little chance of a substantial global price on carbon emissions and there remain relatively high regulatory costs and delays associated with permitting and constructing nuclear power plants which are far higher than those for fossil fuel power facilities. Relatively few new nuclear plants are being commissioned and most are being replaced with cheaper fossil-fuelled alternatives. Nonetheless, uncertainties in the construction time and high capital cost reduce the investment incentive for plants producing low value electricity. We examine how to increase the revenue from nuclear power reactors by more than 100% through co-production of chemicals. Specifically, halogen mediated oxidative dehydrogenation of light alkanes to produce olefins and sulfohalogenation will be described as an exemplars of chemical-nuclear process integration making use the unique features of nuclear reactions to improve the competitiveness of nuclear-based processes. Molten salt-based reactors are particularly well suited for chemical co-production.

IMAGINE THE UNIVERSITY OF UTAH

METALLURGICAL ENGINEERING

Development of Activity Coefficient Model for La(III) in LiCl-KCl Eutectic Salt

Prashant Bagri and Michael Simpson
Department of Metallurgical Engineering
University of Utah

Nuclear Pyrometallurgy Laboratory

IMAGINE THE UNIVERSITY OF UTAH

METALLURGICAL ENGINEERING

Presentation Outline

- Introduction & Theory
- Experimental Set-up
- Results
  - Effect of [$LaCl_3$]
  - Effect of CsCl
  - Effect of Temperature
- Conclusion

Nuclear Pyrometallurgy Laboratory

What is Pyroprocessing?

- High temperature, non-aqueous separation processes applied to spent nuclear fuel.
- Developed in the U.S. under the Integral Fast Reactor program
- Considered one of the top fuel separation technologies for an advanced fuel cycle concept
- Purported to be a closed nuclear fuel cycle

Pyroprocessing Flowsheet

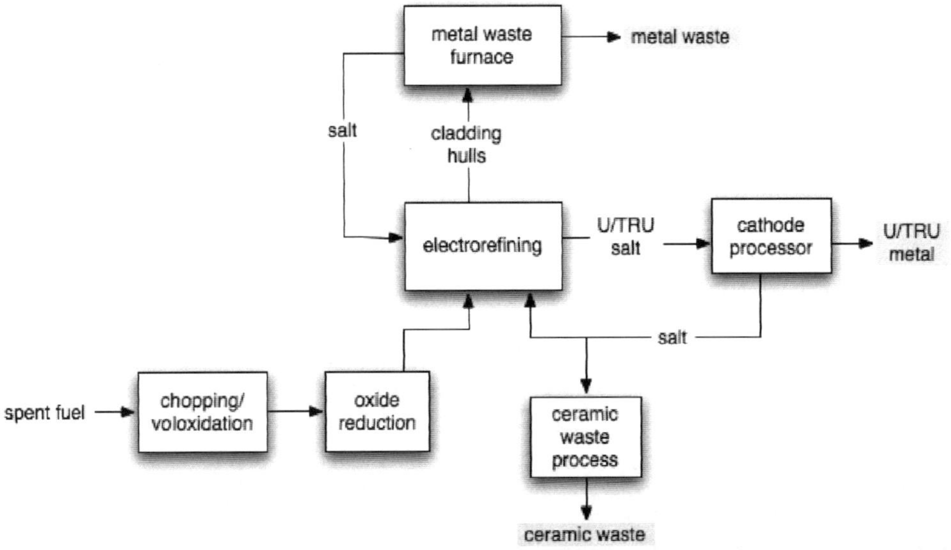

IMAGINE THE UNIVERSITY OF UTAH

METALLURGICAL ENGINEERING

Electrorefining

- Key unit operation in pyro-processing
- Operates with a molten LiCl-KCl electrolyte at 500°C
- Large number of species present presents a challenge.

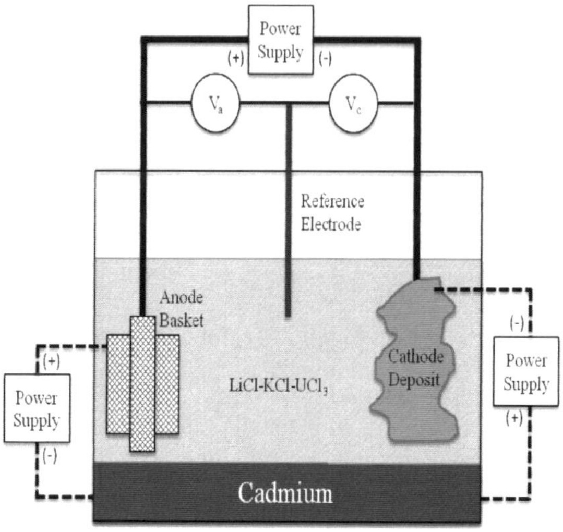

Nuclear Pyrometallurgy Laboratory

IMAGINE THE UNIVERSITY OF UTAH

METALLURGICAL ENGINEERING

Rare Earths (REs) Drawdown

- Use electrolysis for drawdown of REs
- Inert Anode for evolution of Cl_2
- Need to finely control the process requires keen understanding of behavior of complex molten salts (Group I & II elements)
- Accurate modelling of Reduction Potentials in complex molten salts.

Nuclear Pyrometallurgy Laboratory

IMAGINE U THE UNIVERSITY OF UTAH

METALLURGICAL ENGINEERING

Objective

- Gain a more fundamental understanding of the behavior of ions in solution.
- Includes Apparent Potentials and Activity Coefficients.
- Enable enhanced process modelling and control.

Nuclear Pyrometallurgy Laboratory

IMAGINE U THE UNIVERSITY OF UTAH

METALLURGICAL ENGINEERING

Experimental Set-up

- Two Electrode Set-up
- The test was started initially with no $LaCl_3$ present. The $LaCl_3$ was added successively after OCP measurement at every concentration.
- Concentrations determined via ICP-OES.

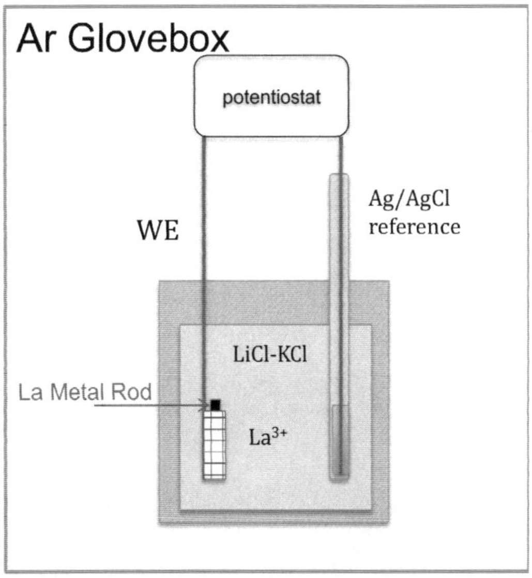

Nuclear Pyrometallurgy Laboratory

IMAGINE THE UNIVERSITY OF UTAH

METALLURGICAL ENGINEERING

Nernst Equation

For any elementary reaction,

$$M^{n+} + ne^{-} = M^{0}$$

The Nernst Equation is as follows:

$$E^{eq} = E^{0} - \frac{RT}{nF} \ln\left(\frac{a_{M^0}}{a_{M^{n+}}}\right)$$

Which can be re-written as:

$$E^{eq} = E^{0} + \frac{RT}{nF} \ln\left(\gamma_{M^{n+}} X_{M^{n+}}\right)$$

Since,
$$a_{M^{n+}} = \gamma_{M^{n+}} X_{M^{n+}}$$

Re-arranging

$$E^{eq} = E^{0} + \frac{RT}{nF} \ln\left(\gamma_{M^{n+}}\right) + \frac{RT}{nF} \ln\left(X_{M^{n+}}\right)$$

Making adequate substations,

$$\ln\left(\gamma_{M^{n+}}\right) = \frac{\Delta G^{SC} - \Delta G^{OCP} - RT\ln\left(X_{M^{n+}}\right)}{RT}$$

$$RT * lna(i) = [G(i) - G(ref)]$$

The activity (a) is a relative property. We define the reference state as "Pure Supercooled Liquid $LaCl_3$"

Nuclear Pyrometallurgy Laboratory

IMAGINE THE UNIVERSITY OF UTAH

METALLURGICAL ENGINEERING

Results

- Effect of [$LaCl_3$] on γ_{LaCl3} of $LaCl_3$
  - T = 500°C in a x$LaCl_3$-LiCl-KCl solution.
- The presence of other ions viz. CsCl on activity coefficient (γ_{LaCl3}) of $LaCl_3$.
  - T = 500°C in a x$LaCl_3$-LiCl-KCl-2 wt. % CsCl solution.
- Effect of Temperature on γ_{LaCl3}

Nuclear Pyrometallurgy Laboratory

IMAGINE THE UNIVERSITY OF UTAH

METALLURGICAL ENGINEERING

OCP Measurement vs. $[LaCl_3]$

- $La(III)/La^0$ equilibrium measurement
- A model was fit to the OCP data measured.
- $\Delta E < 5mV$
- These OCP^{model} values were used for subsequent calculations

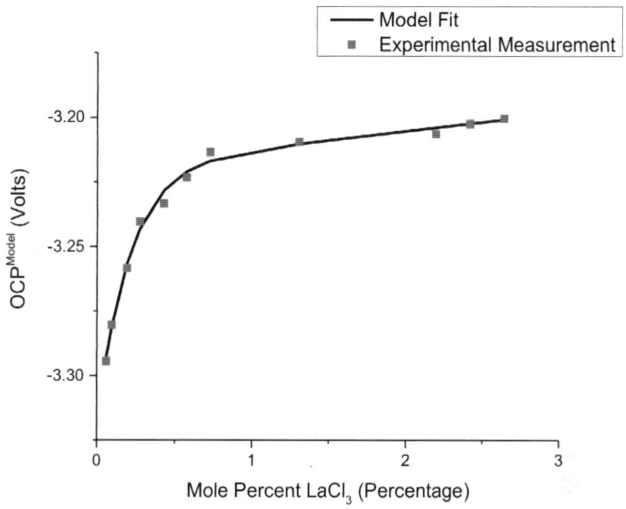

Nuclear Pyrometallurgy Laboratory

IMAGINE THE UNIVERSITY OF UTAH

METALLURGICAL ENGINEERING

Apparent Potential ($E^{0'}$) of La(III) vs. $[LaCl_3]$

- The Apparent Potential value is calculated using the Nernst Equation as follows:

$$E^{0'} = E^0 + \frac{RT}{nF} ln\gamma = E(eq.) - \frac{RT}{nF} lnX_{\text{La}}$$

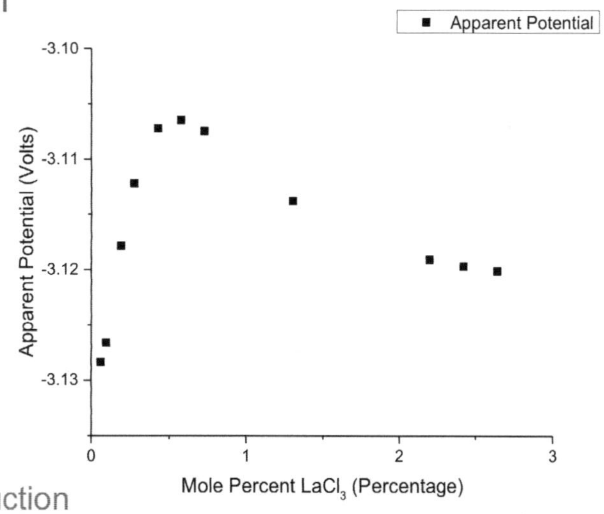

25 mV window for La(III) reduction

Nuclear Pyrometallurgy Laboratory

Activity Coefficient (γ_{LaCl3}) vs. [$LaCl_3$]
Excess Free Energy (ΔG^E) vs. [$LaCl_3$]

- Data shows γ_{LaCl3} varies with concentration!!

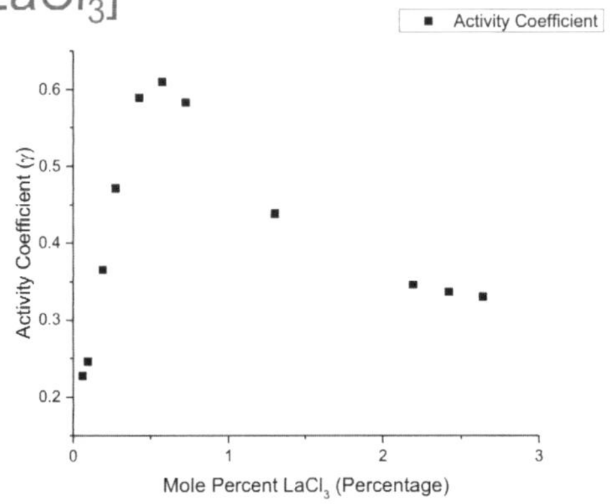

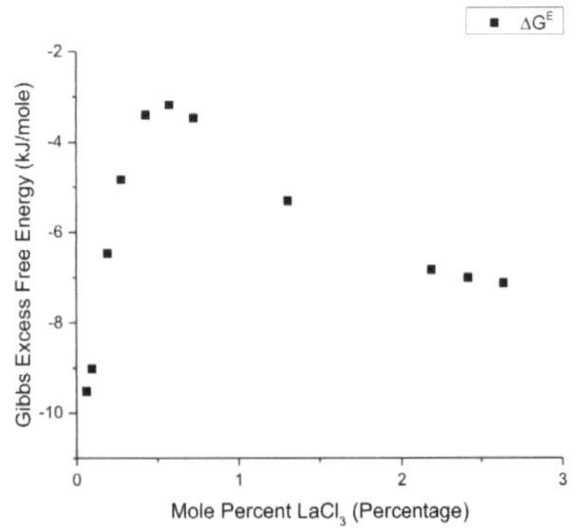

- ΔG^E is the Excess Gibbs free energy and is synonymous to activity coefficient:

$$G^E = RT ln\gamma$$

Results

- Effect of [$LaCl_3$] on γ_{LaCl3} of $LaCl_3$
  - T = 500°C in a $xLaCl_3$-LiCl-KCl solution.
- The presence of other ions viz. CsCl on activity coefficient (γ_{LaCl3}) of $LaCl_3$.
  - T = 500°C in a $xLaCl_3$-LiCl-KCl-2 wt. % CsCl solution.
- Effect of Temperature on γ_{LaCl3}

IMAGINE THE UNIVERSITY OF UTAH

METALLURGICAL ENGINEERING

OCP Measurement vs. [$LaCl_3$]

- La(III)/La^0 equilibrium measurement.
- $\Delta E < 2mV$

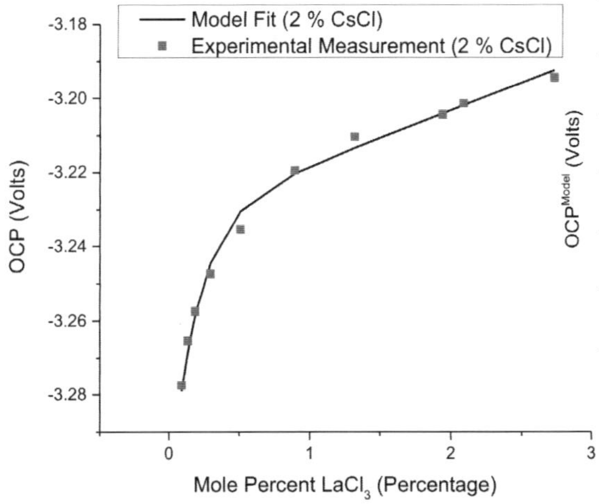

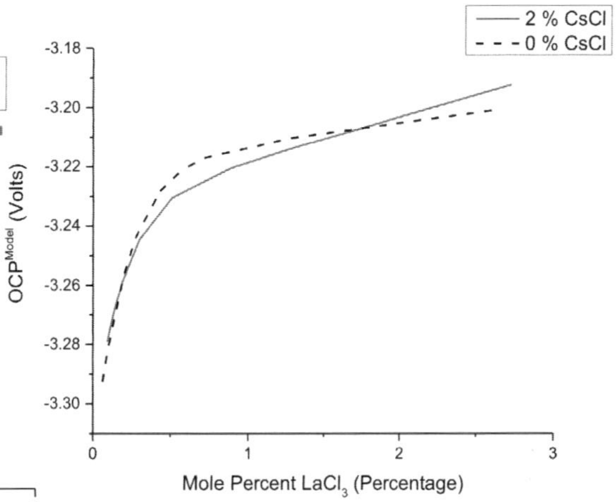

Nuclear Pyrometallurgy Laboratory

IMAGINE THE UNIVERSITY OF UTAH

METALLURGICAL ENGINEERING

Apparent Potential ($E^{0'}$) of La(III) vs. [$LaCl_3$]

CsCl reduces window for La(III) reduction!!

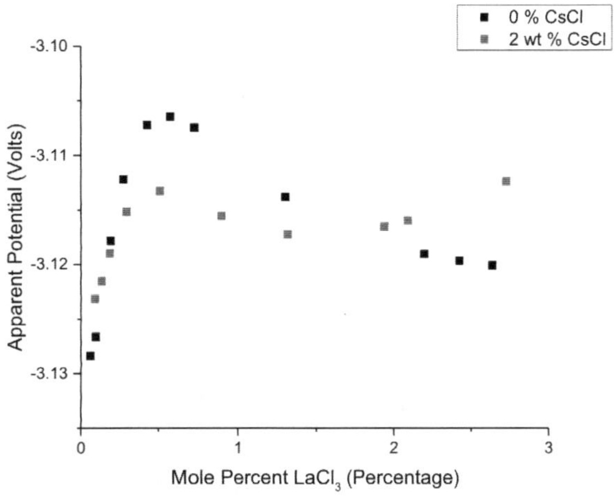

Nuclear Pyrometallurgy Laboratory

IMAGINE THE UNIVERSITY OF UTAH

METALLURGICAL ENGINEERING

Activity Coefficient (γ_{LaCl3}) vs. [$LaCl_3$] Excess Free Energy (ΔG^E) vs. [$LaCl_3$]

- Presence of CsCl seems to affect the activity of La ions in solution

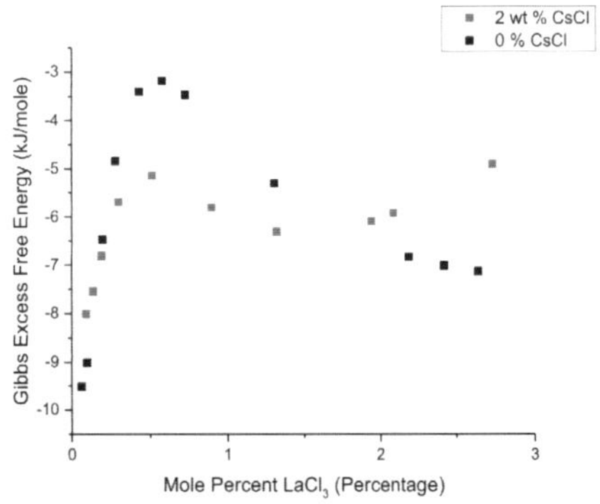

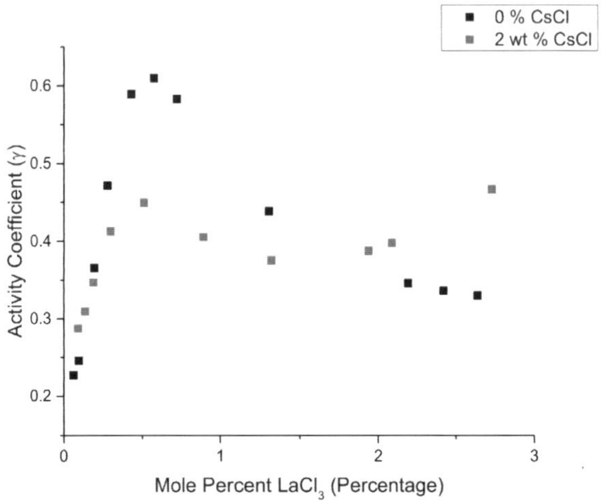

Nuclear Pyrometallurgy Laboratory

IMAGINE THE UNIVERSITY OF UTAH

METALLURGICAL ENGINEERING

Activity of La(III) ions

- Ideal Behavior ➔ $\gamma = 1$
- Significant Deviation observed!!!

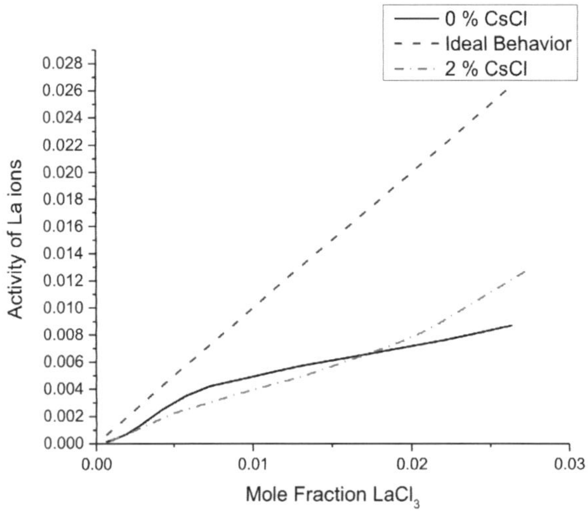

Nuclear Pyrometallurgy Laboratory

IMAGINE THE UNIVERSITY OF UTAH

METALLURGICAL ENGINEERING

Results

- Effect of [$LaCl_3$] on γ_{LaCl3} of $LaCl_3$
  - T = 500°C in a $xLaCl_3$-LiCl-KCl solution.
- The presence of other ions viz. CsCl on activity coefficient (γ_{LaCl3}) of $LaCl_3$.
  - T = 500°C in a $xLaCl_3$-LiCl-KCl-2 wt. % CsCl solution.
- Effect of Temperature on γ_{LaCl3}

Nuclear Pyrometallurgy Laboratory

IMAGINE THE UNIVERSITY OF UTAH

METALLURGICAL ENGINEERING

Effect of Temperature on Activity Coefficient

- Lower [$LaCl_3$] ➔ Higher γ
- T = 400°C, 425°C, 450°C & 500°C

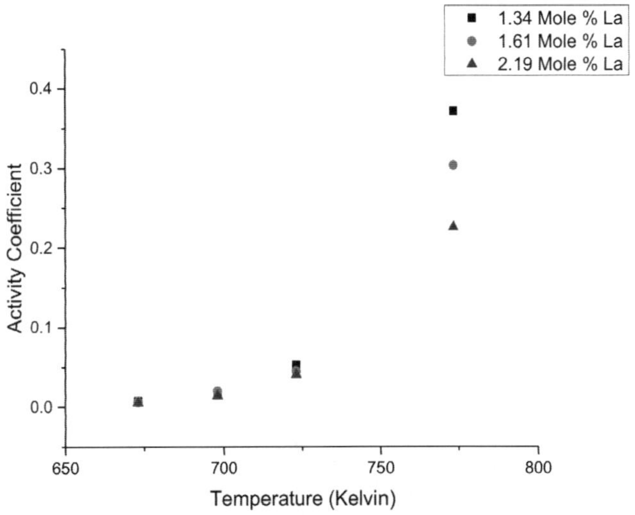

Nuclear Pyrometallurgy Laboratory

IMAGINE THE UNIVERSITY OF UTAH

METALLURGICAL ENGINEERING

Conclusions

- New data reported here provides evidence that the activity coefficient of $LaCl_3$ depends on the solution matrix & temperature
- The presence of CsCl affects the activity of $LaCl_3$.

Nuclear Pyrometallurgy Laboratory

IMAGINE THE UNIVERSITY OF UTAH

METALLURGICAL ENGINEERING

Acknowledgements

- This work was supported by the US Department of Energy's Nuclear Energy University Program (NEUP) Project Number NEUP-14-6489

Nuclear Pyrometallurgy Laboratory

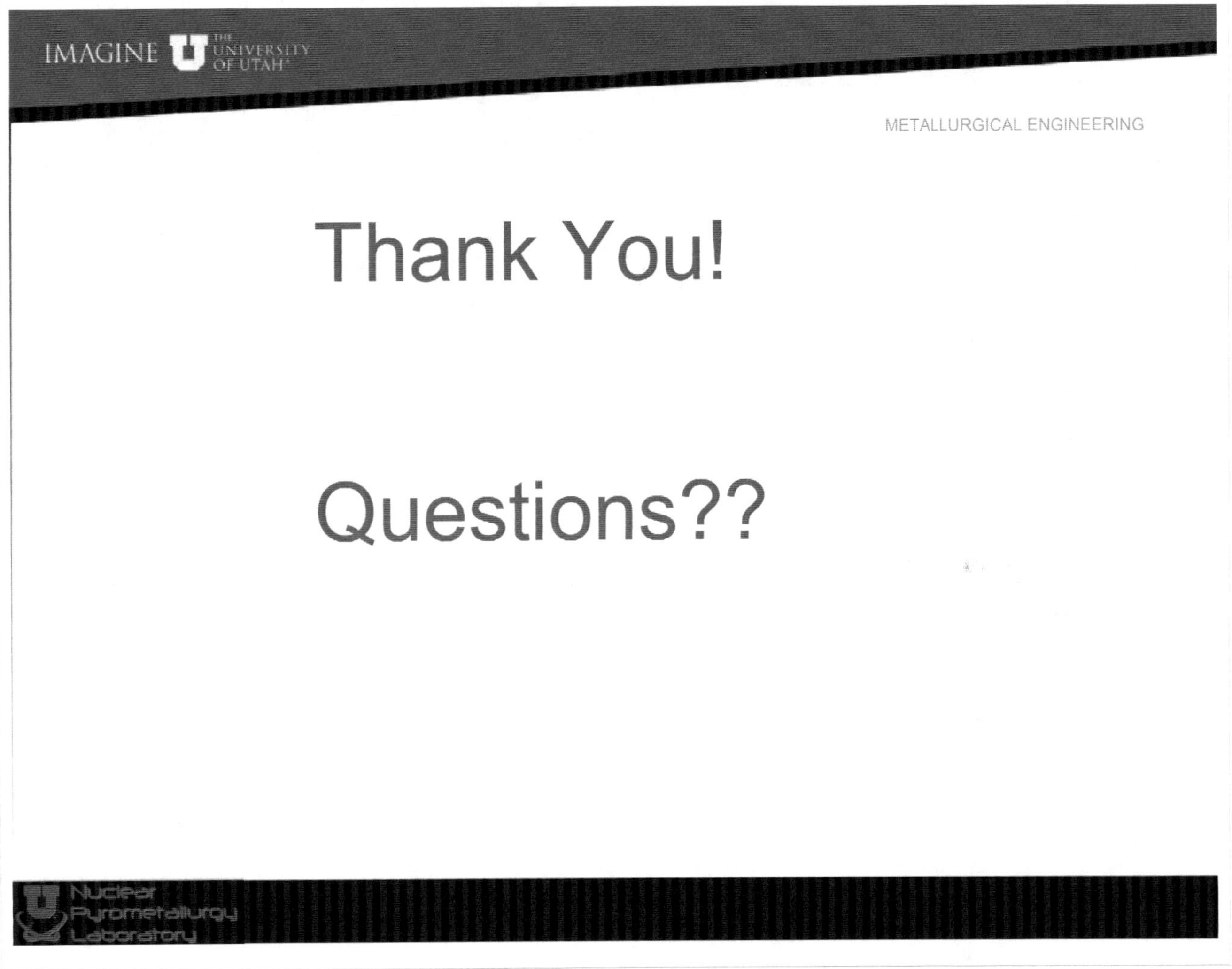
IMAGINE U THE UNIVERSITY OF UTAH
METALLURGICAL ENGINEERING
Thank You!
Questions??
Nuclear
Pyrometallurgy
Laboratory

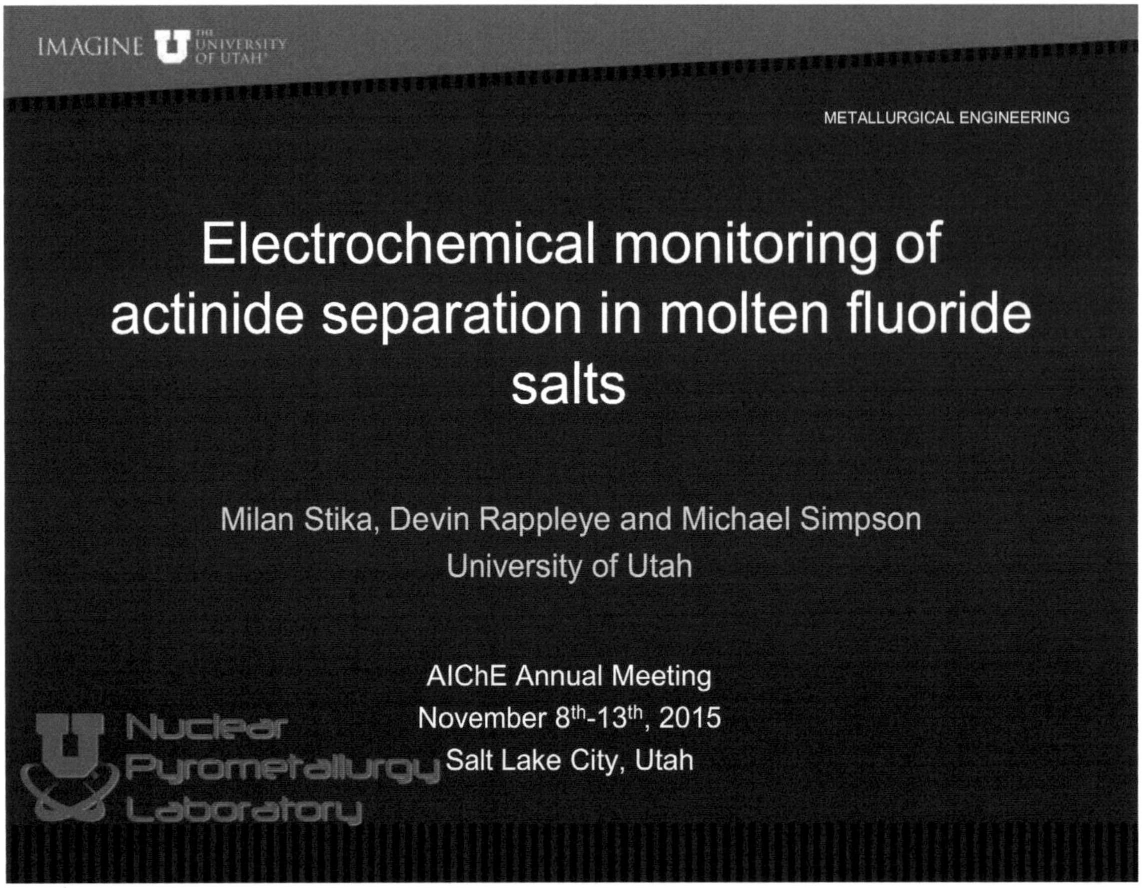
IMAGINE THE UNIVERSITY OF UTAH
METALLURGICAL ENGINEERING
Electrochemical monitoring of actinide separation in molten fluoride salts
Milan Stika, Devin Rappleye and Michael Simpson
University of Utah
AIChE Annual Meeting
November 8th-13th, 2015
Salt Lake City, Utah
Nuclear Pyrometallurgy Laboratory

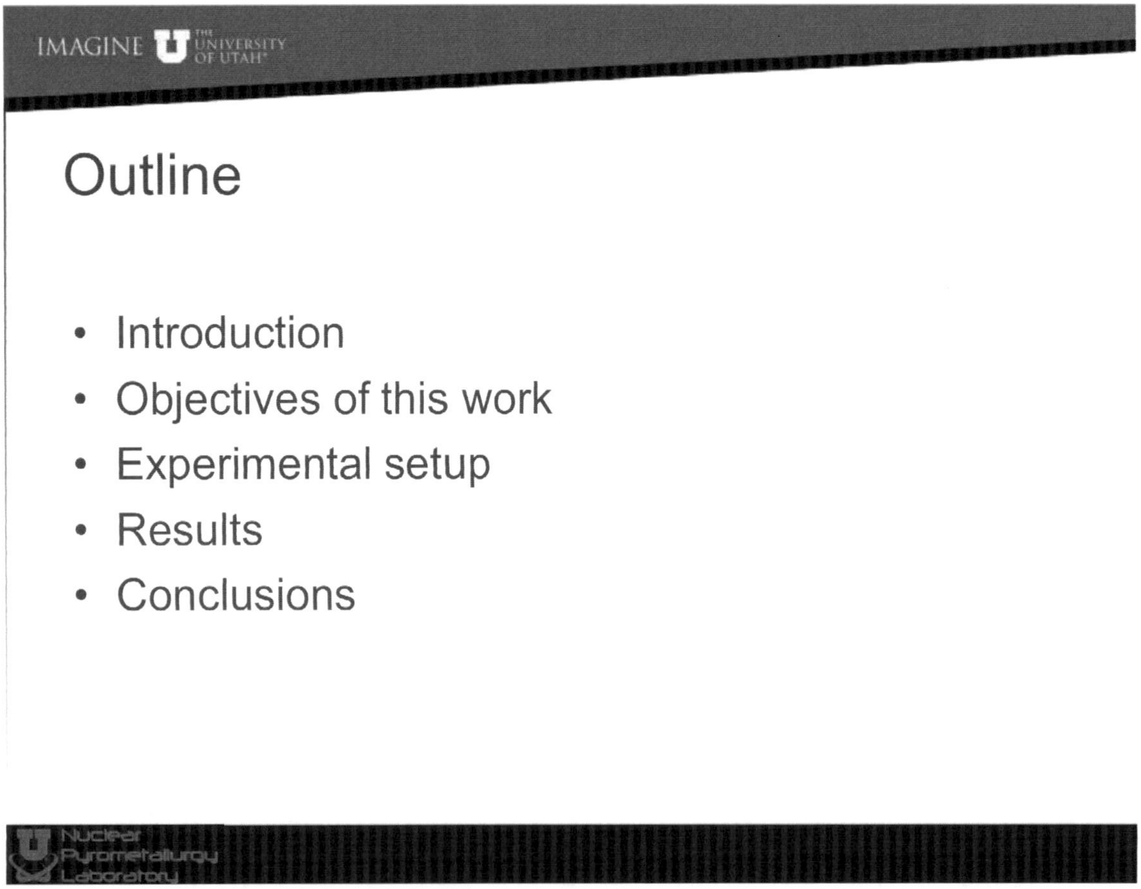
IMAGINE THE UNIVERSITY OF UTAH
Outline
• Introduction
• Objectives of this work
• Experimental setup
• Results
• Conclusions
Nuclear Pyrometallurgy Laboratory

Introduction

- Making nuclear energy sustainable in the long run
- Thorium-uranium fuel cycle harbors significant advantages over traditional uranium-plutonium fuel cycle:
- Minimization of transuranic long-lived HLW
- ^{233}U fissions better than ^{235}U
- Potential for elimination of some costly steps (enrichment, fuel fabrication) => economical advantage
- Proliferation resistance – difficult manipulation with ^{233}U, generally unsuitable for weaponization
- All these benefits best utilized in a molten salt reactor which uses fluoride fuel

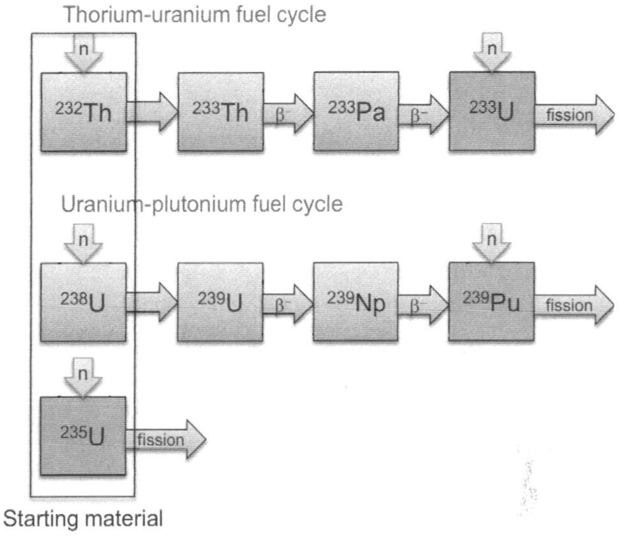

IMAGINE U THE UNIVERSITY OF UTAH

Introduction II

- Reactor salt: LiF-BeF_2-UF_4
- Blanket salt: LiF-BeF_2-ThF_4
- Focus on blanket salt imitation: LiF-ThF_4
- Blanket salt breeds ^{233}U via ^{232}Th
- Salt is periodically processed to extract bred ^{233}U (and ^{233}Pa)
- ORNL method of choice: Salt/Bi extraction
- Novel method: Electrodeposition of U on inert cathode

$$^{232}Th + n \rightarrow {}^{233}Pa \xrightarrow{T_{1/2}=27d} {}^{233}U$$

$$^{233}Pa + n \rightarrow {}^{234}Pa \rightarrow {}^{234}U$$

Nuclear Pyrometallurgy Laboratory

IMAGINE THE UNIVERSITY OF UTAH

Objectives of this work

- Select suitable containers, electrodes, etc. for molten fluoride/molten bismuth experiments
- Obtain concentration "calibration curve" (reduction peak current height vs. analyte concentration) for Th and U using cyclic voltammetry (CV)
- Use these in a) salt/Bi extraction experiments to track the progress of extraction
- Estimate extraction feasibility
- Use these in b) electrodeposition experiments to track the progress of deposition
- Estimate electrodeposition feasibility

Nuclear Pyrometallurgy Laboratory

IMAGINE THE UNIVERSITY OF UTAH

Experimental setup

- 3 electrode setup: Mo (inert WE), glassy carbon or SS basket (CE), Pt (quasi-RE)
- Glassy carbon, Mo crucibles
- Glove box with UHP argon atmosphere (< 1 ppm O_2, H_2O)
- Kerr furnace operated at 900 C (1173 K)
- 99.98% LiF (Alfa Aesar), dried in-house
- 99.99% ThF_4 (Materion)
- 99.8% UF_4 (IBI-Labs)
- 99.99% Bi needles (Alfa Aesar)
- Th turnings
- CVs (100-300 mV/s) performed using Metrohm Autolab potentiostat with NOVA software
- Salt content analyzed by ICP-MS

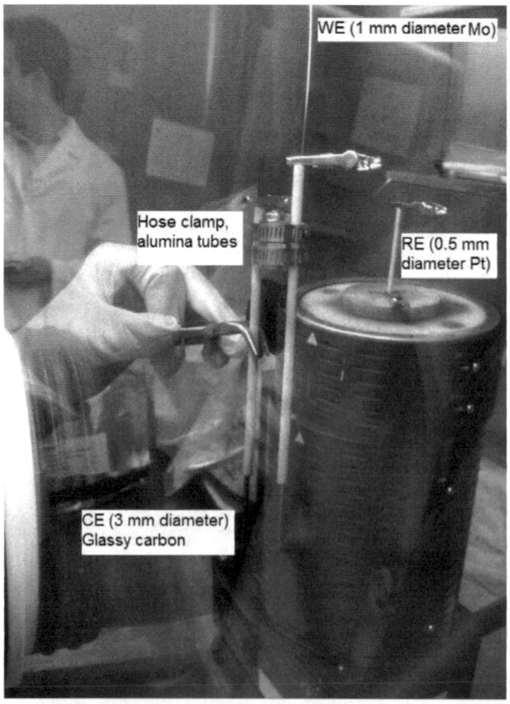

Nuclear Pyrometallurgy Laboratory

IMAGINE U THE UNIVERSITY OF UTAH

Salt/Bi extraction experiments

- Reductive extraction of U into Bi:

$$UF_4 + 4Li(Bi) \rightarrow U(Bi) + 4LiF$$

- How to prepare reproducible salt/Bi system?
- A1) Melt Bi needles, cool down to get Bi ingot
- A2) Melt ingot, add known amount of Li into Bi to get Bi-Li ingot (Li is quite soluble in Bi)
- B1) Dry LiF salt to eliminate moisture
- B2) Melt LiF-ThF_4 salt
- B3) Realize CV measurement on the salt to verify electrochemical purity and ThF_4 conc., cool down to get salt ingot
- C1) Combine Bi-Li ingot and salt ingot together, melt
- C2) Extraction commences immediately after both phases are molten (you can't capture start)
- C3) Try to capture some extraction progress with electrochemical means
- C4) Meanwhile, periodically sample the salt for later chemical analysis
- C5) End experiment, hope your crucible does not crack upon cooling

Nuclear Pyrometallurgy Laboratory

IMAGINE U THE UNIVERSITY OF UTAH

Salt/Bi extraction experiments II

- Both phases contained in a single GC crucible
- No undesirable salt/Bi or crucible/Bi chemical reactions expected
- Mechanical constraints
- Salt shrinks during solidification (~760 C) – usually easy removal from GC crucible
- Bi expands during solidification (~270 C) = analogy to water/ice anomaly. Crucible will crack or the frozen salt will be pushed upwards => use tapered crucible!
- Friction prevents the removal of Bi from crucible => use tapered crucible!
- Solid salt/solid Bi interface is not easily separable (use bigger hammer)
- This prevents the removal of salt only out of crucible (you might want to re-use Bi)

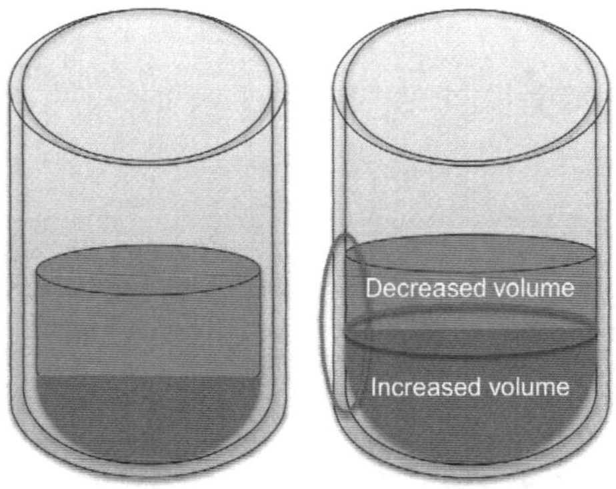

Molten system Frozen system

Nuclear Pyrometallurgy Laboratory

IMAGINE THE UNIVERSITY OF UTAH

Salt/Bi extraction experiments III

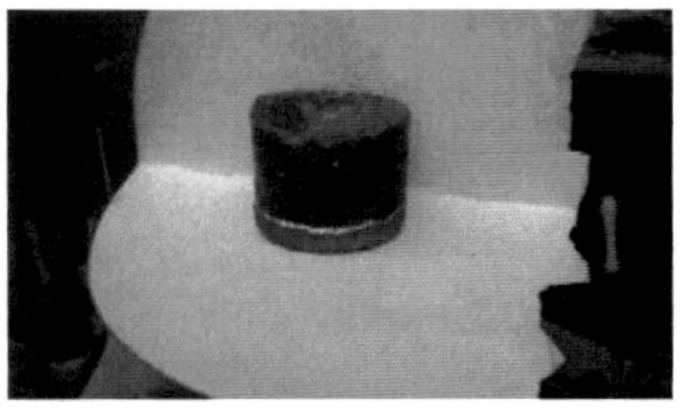

- Several cracked crucibles later:

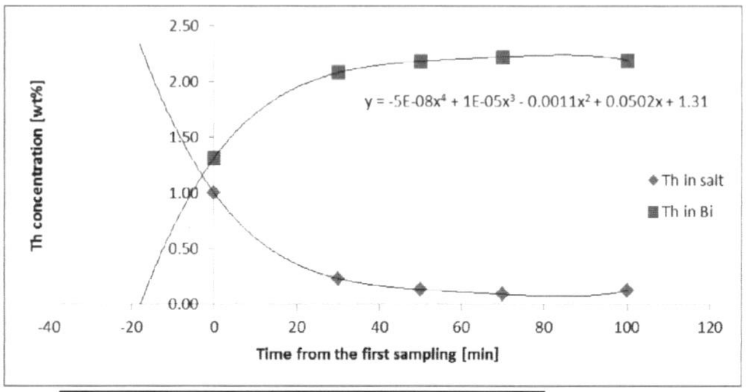

Actual time	Time [min]	Th in salt [wt%] (init.: 2.31 wt%)	Th in Bi [wt%] (complement)
2:50	0	1.00	1.31
3:20	30	0.23	2.08
3:40	50	0.13	2.18
4:00	70	0.09	2.22
4:30	100	0.12	2.19

Nuclear Pyrometallurgy Laboratory

IMAGINE THE UNIVERSITY OF UTAH

Salt/Bi extraction experiments IV

- Goal was to monitor the progress electrochemically (be able to draw a similar extraction curve based on elchem. data)
- Previous work produced calibration curve: peak current density = f(concentration)

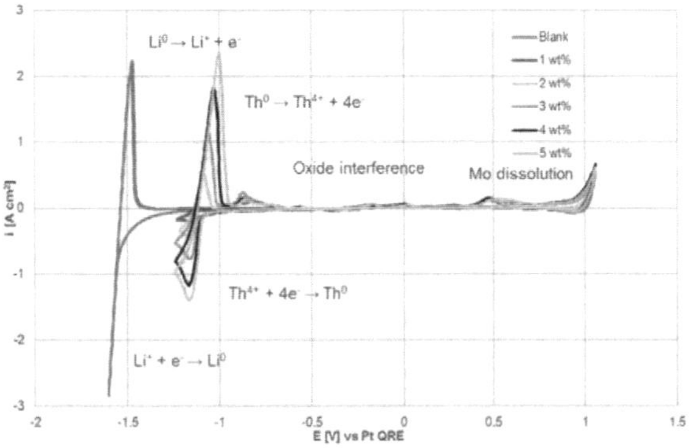

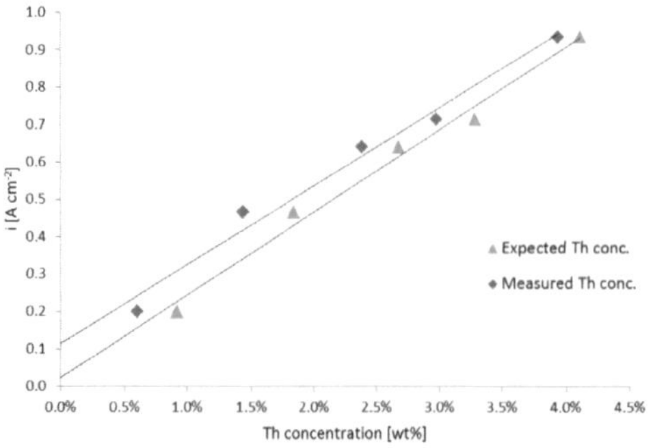

Nuclear Pyrometallurgy Laboratory

IMAGINE U THE UNIVERSITY OF UTAH

Salt/Bi extraction experiments V

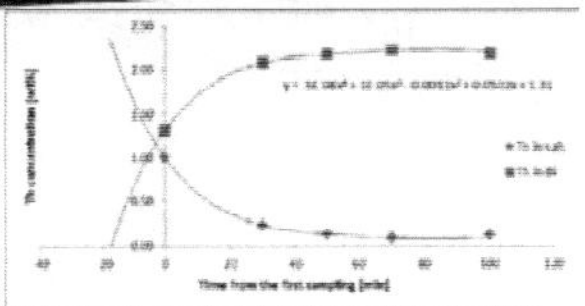

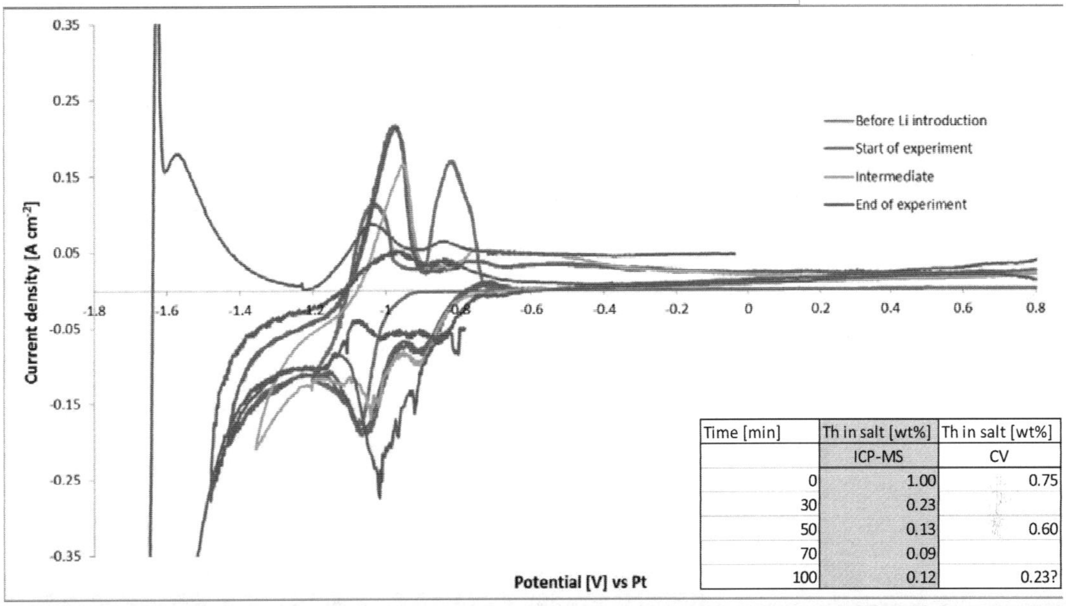

Time [min]	Th in salt [wt%]	Th in salt [wt%]
	ICP-MS	CV
0	1.00	0.75
30	0.23	
50	0.13	0.60
70	0.09	
100	0.12	0.23?

Nuclear Pyrometallurgy Laboratory

IMAGINE U THE UNIVERSITY OF UTAH

Salt/Bi extraction experiments VI

- Conclusion:
- Tricky experimental setup, better to have two separate crucibles
- Due to this: the beginning of the extraction (most drastic change in concentration) can never be captured
- Due to this: electrochemical measurement not optimal, Bi entrained in the salt causes problems, weird current vs. potential response
- Problematic to obtain reliable data for concentration monitoring

- Verified that Th readily extracts into Bi (bad for U/Th separation in the blanket, since U extracts as well)
- Very important to add just enough reductant to preferentially reduce U and not Th (but we haven't got that far in this study)
- => maybe other U/Th separation technique is more accessible for investigation

Nuclear Pyrometallurgy Laboratory

IMAGINE U THE UNIVERSITY OF UTAH

Electrodeposition of U

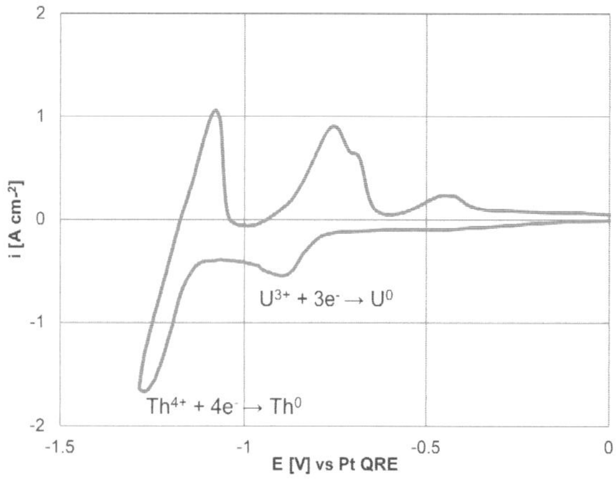

- Previous work with $LiF\text{-}ThF_4\text{-}UF_4$
- Good distance between U and Th reduction potentials reveals practical separation possible (unlike chloride systems)
- U is the first element to reduce

- Cathode morphology drastically changes during U deposition, possible dendrite growth – unpredictable rise in surface area => prevents reliable concentration measurement of Th (current density value is larger due to much larger surface area)

Nuclear Pyrometallurgy Laboratory

IMAGINE U THE UNIVERSITY OF UTAH

Electrodeposition of U II

- Electrodeposition under constant potential = chronoamperometry (CA)
- CA is electrochemical process, obvious choice to monitor concentration electrochemically as well
- Previous work produced calibration curve: peak current density = f(concentration)

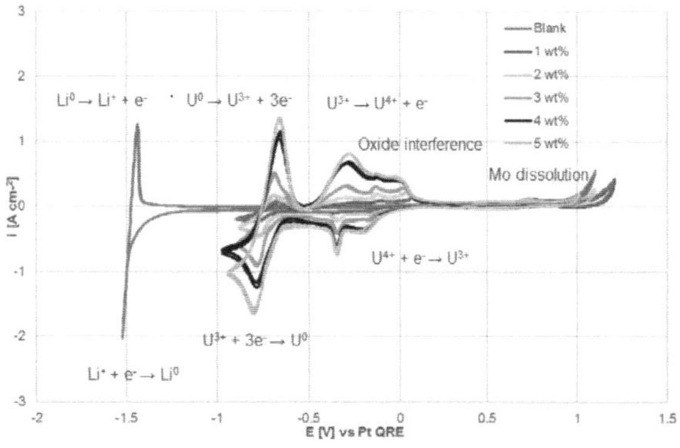

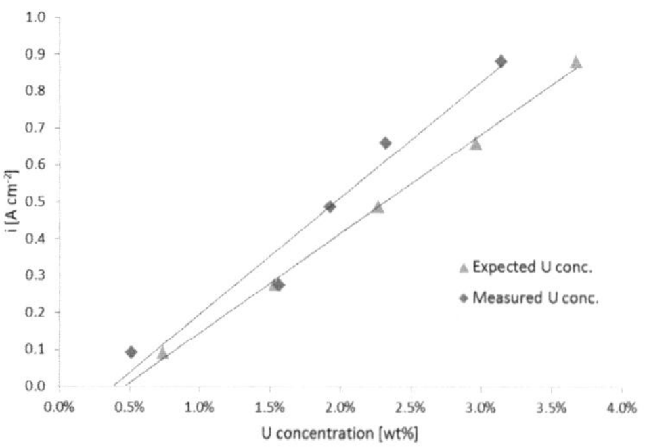

Nuclear Pyrometallurgy Laboratory

IMAGINE U THE UNIVERSITY OF UTAH

Electrodeposition of U III

- Experimental setup
- LiF-UF_4 (2.6 wt% U)
- Mo WE (cathode):

$$U^{4+} + e^- \rightarrow U^{3+}$$
$$U^{3+} + 3e^- \rightarrow U$$

- Pt RE
- Th turnings inside SS basket as CE (anode):

$$Th \rightarrow Th^{4+} + 4e^-$$

- Th turnings oxidize to ThF_4 which should be detectable at the end of CA run
- Pass known amount of charge to reduce significant amount of UF_4 into U metal

$$Q = \int I\,dt$$

$$Q = \frac{m}{M} nF$$

Nuclear Pyrometallurgy Laboratory

IMAGINE U THE UNIVERSITY OF UTAH

Electrodepositi

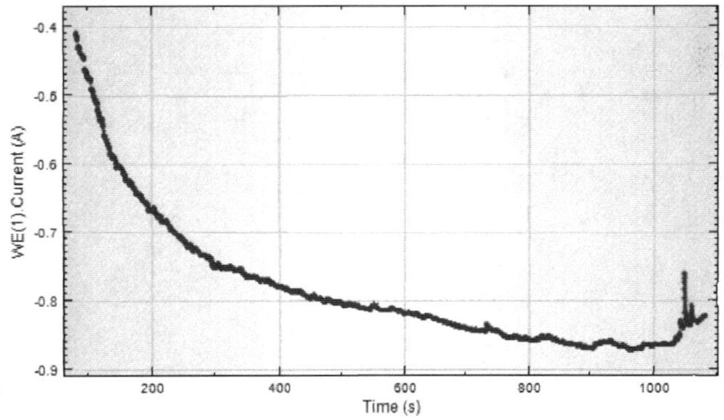

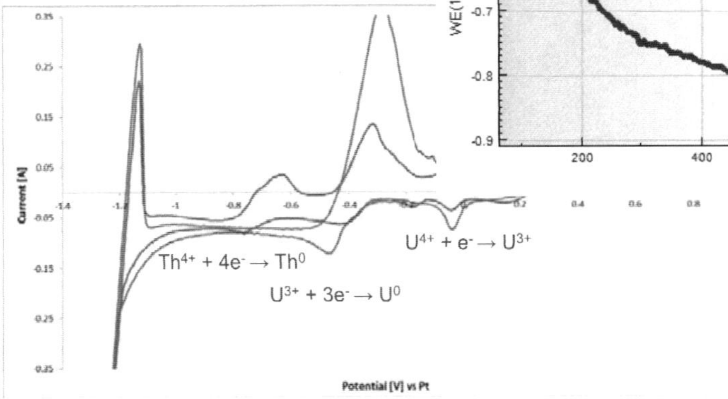

	BEFORE		AFTER		Difference	
	ICP-MS	CV	ICP-MS	CV	ICP-MS	CV
U in salt [wt%]	2.56	2.19	1.59	1.08	-0.97	-1.11
Th in salt [wt%]	0.27	0	1.75	?	1.48	?

lower than U reduction potential (around -0.6 V)

- Current was steadily increasing during the electrodeposition => surface area was growing
- Theoretical U reduction: 38 % (1200/3180)
- Actual reduction: 38 % (ICP-MS)
- Actual reduction: 51 % (CV)

Nuclear Pyrometallurgy Laboratory

IMAGINE THE UNIVERSITY OF UTAH

Conclusions

- Electrodeposition of U on inert electrode was successful
- Theoretical amount of U reduced (38 %) was exactly matched by ICP-MS results
- Theoretical amount of U reduced calculated from CV was moderately overestimated (51 %)
- Overestimation might be partially due to imperfection in concentration calibration curve developed earlier, simple after/before peak ratio gives 47 % U reduction
- Th peak was identified after electrodeposition, concentration was not calculated due to known problem of severe overestimation encountered in earlier work (only first reduced species on the electrode can be reliably measured)
- No major problems compared to previous investigation were encountered

Nuclear Pyrometallurgy Laboratory

IMAGINE THE UNIVERSITY OF UTAH

Thank you

Nuclear Pyrometallurgy Laboratory

420190 Measurement of Electrochemical Transport Properties of Manganese Ions in Eutectic LiCl-KCl Using High Temperature Voltammetry

Abstract:

Measurement of Electrochemical Transport Properties of Manganese Ions in Eutectic LiCl-KCl using High Temperature Voltammetry

D. Horvath and M.F. Simpson

University of Utah, Department of Metallurgical Engineering

A study of electrochemical deposition of manganese from molten salt mixtures of LiCl-KCl-$MnCl_2$ was carried out using a variety of voltammetry methods in a high temperature electrochemical test system. Deposition was studied on both tungsten and molybdenum working electrodes in electrodes in eutectic LiCl-KCl melts at 773 K. This study was motivated by a need to minimize transition metal impurity concentrations in electrorefined uranium metal. Though manganese is electrochemically nobler then uranium, its standard reduction potential is relatively close to that of U. And it is present in the stainless steel hardware that is immersed in the molten salt in an electrorefiner along with spent fuel or other sources of impure uranium. Study of manganese is expected to provide insight to the behavior of a number of different transition metals that may contaminate the uranium, including zirconium, chromium, and iron. For this study, transient electrochemical techniques such as cyclic voltammetry, chronopotentiometry, and chronoamperometry were used to investigate the deposition behavior of Mn^{2+} ions from LiCl-KCl with varying concentrations of $MnCl_2$. Molybdenum was found to be the preferred working electrode, as double anodic peaks were measured for cyclic voltammetry using tungsten electrodes. This has been attributed to possible formation of an intermetallic W-Mn compound. Diffusion coefficient, exchange current density, and other properties have been evaluated at $MnCl_2$ concentrations ranging from 0.05 to 0.75wt%. Even over this relatively short range, concentration was found to affect these properties. Diffusion coefficient calculations also varied depending on which voltammetry method was used for the measurement. Cyclic voltammetry tests yielded D values from 6.8×10^{-6} to 1.8×10^{-5} cm^2/sec. Standard exchange current density was likewise determined using CV based on the slope of the reducing peak near the apparent reduction potential--ranging from 0.18 to 0.58 A/cm^2. These properties can ultimately be used to model the electrorefining process, estimate the extent of transition metal contamination of uranium, and provide insight in how to minimize this contamination.

433017 A Combined Microstructural and Chemical Study of Uranium in Unconventional Shale Reservoirs

Abstract:

A Combined Microstructural and Chemical Study of Uranium in Unconventional Shale Reservoirs

Joseph D. Levinthal[1*], Bryony Richardds-McClung[2], and Luther McDonald[1]

[1]University of Utah Nuclear Engineering Program

[2] Energy Geoscience Institute

Abstract: Unconventional 'shale' reservoirs have moved from a minority interest to center stage focus, and are being evaluated for their potentially large hydrocarbon (oil and/-or gas) reserves. During the hydraulic fracturing of these shale formations, many additional minerals are solubilized and brought to the surface, or potentially transported to ground water, which has generated some concerns over the safety of these techniques. This study focuses on the morphological properties of uranium, which were investigated to develop a knowledgeable interpretation of environments of deposition and uranium mineralogical associations within the shale. This knowledge leads to a more advanced understanding of the sedimentological evolution of micro- and macro-environments, and is key to the reconstruction of historical basins, oceans, landscapes, climates, climate cycles, and enables the future development of techniques that may help reduce the migration of radionuclides in these environments. It additionally supports the analysis of the feasibility of selectively extracting the uranium that has been brought to the surface with proppant, and associated hydraulic fracturing chemicals. The mineralogical and chemical evaluation of the shale samples was investigated by X-ray Diffraction (XRD), Quantitative Evaluation Of Minerals By Scanning Electron Microscopy (QEMSCAN®), and by Scanning Electron Microscopy/Energy Dispersive X-ray Spectroscopy (SEM/EDX). Radiochemical separations were performed to extract the uranium and inductively coupled plasma mass spectrometry (ICP-MS) was used to quantify the uranium concentration for the samples.

411797 Evaporation Tests Using Redc Lllw Simulant Solutions

Abstract:

EVAPORATION TESTS USING REDC LLLW SIMULANT SOLUTIONS

P. A. Taylor

Oak Ridge National Laboratory

The Radiochemical Engineering Development Center (REDC) at Oak Ridge National Laboratory (ORNL) is a hot cell facility that recovers heavy elements from irradiated targets and produces targets for irradiation. Since the mid-1960s, REDC has been the production, storage, and distribution center for transuranium elements used for the heavy-element research program of the US Department of Energy and other users. REDC discharges about 45,000 L/year of liquid low-level waste (LLLW), mostly aqueous with 2–3% organics, into the ORNL central LLLW system. The total LLLW generation rate has been about 700,000 L/year, with most of the volume generated by environmental management (EM)-related activities. REDC is the primary source of radionuclides discharged to the LLLW system. LLLW is evaporated, and the concentrate—a mixture of sludge and supernate—is stored in underground tanks. The Transuranic Waste Processing Center (TWPC) has solidified and disposed of most of the supernate and will treat the sludge in the future. Remediation work eventually will eliminate most of the EM-generated LLLW. The existing LLLW treatment system and TWPC are planned for decontamination and decommissioning after the LLLW generation rate is reduced. The remaining LLLW will contain a much higher radionuclide concentration, and the chemical composition will change. New evaporation and solidification facilities will be needed to treat future LLLW.

A series of laboratory-scale evaporation tests have been conducted using a simulant of the LLLW generated at REDC. The purpose of these tests was to determine the volume reduction factor (VRF = initial volume / final volume) that can be achieved before solids start to form, the VRF where the evaporator concentrate solidifies into a salt cake, and the distribution of the organics between the condensate and concentrate solutions. The tests also were intended to identify potential operational issues with evaporating this solution. A simulant solution was formulated from 11 years (2001–12) of data on the composition of LLLW discharges from REDC (see Table 1).

Table 1. Aqueous LLLW Simulant Composition

Chemical	**Concentration (*M*)**
Na_2CO_3	0.13
NaOH	0.19
$NaNO_3$	0.41
NaCl	0.15
$NaAlO_2$	0.02–0.6

The sodium aluminate ($NaAlO_2$) concentration has averaged 0.02 *M*, but future programs may increase the concentration, so concentrations of 0.1 and 0.6 *M* also were tested. The organic portion of LLLW includes extractants, such as 1,4-diethylbenzene, tributyl phosphate, and Bis(2 ethylhexyl) hydrogen phosphate, as well as petroleum naphtha diluents.

Solids started to form in the boiling simulants at a VRF above 5.2 for the simulant with 0.02 *M* $NaAlO_2$, at a VRF = 2.9 for the 0.1 *M* $NaAlO_2$ simulant, and at a VRF = 2.1 for the 0.6 *M* $NaAlO_2$ simulant. Solids formed at lower VRFs when the solution was cooled. Most of the organic diluent evaporated, but the extractants stayed with the concentrated simulant and started to solidify after being boiled for many hours.

Aspen/OLI Model of Waste Evaporation and Neutralization Process for MOX Waste Solidification Building

Dr. James E. Laurinat, Savannah River National Laboratory, Aiken, SC

Introduction

The Mixed Oxide Fuel Fabrication Facility (MOX or MFFF) at the Savannah River Site includes a Waste Solidification Building (WSB), which was built to house the process for converting aqueous wastes from the MFFF and the proposed Pit Disassembly and Conversion Facility (PDCF) into cements for long-term storage in either onsite repositories or at the Waste Isolation Pilot Plant (WIPP). The WSB process design required flowsheet modeling of the mixing tanks, batch evaporators, and neutralization tanks included in the WSB process.

A steady state computer simulation was developed, using Aspen Plus® to model the mixing tanks and evaporators and OLI® to model the chemistry. The simulation is configured so that both inputs and outputs appear on a Microsoft Excel® spreadsheet.

Aspen Model Details

Simulations for expected and maximum flow rates and acid and salt concentrations were requested for characteristic MOX feeds from the Pit Disassembly and Conversion Facility (PDCF) and from the Alternate Feed Stock (AFS). Each set of inputs includes specifications for three High Activity Waste (HAW) streams, a stream containing significant concentrations of Am-241 (AM241 in the simulation), an alkaline stream (ALKALINE), and an excess concentrated nitric acid stream (XSACID), and two Low Activity Waste (LAW) streams, a stream carrying the PDCF laboratory wastes (PDCFLAB), and a strip solution stream with a high concentration of primarily depleted uranium (STRIPU).

All input stream compositions are defined in terms of a list of chemical basis species, which comprise the OLI chemistry model. All input species are required to be neutral acids, salts, or hydroxides included in the OLI databooks. Each chemical element corresponds to one chemical basis species in the OLI chemistry model, with a few exceptions. The primary exception is sodium (Na), which is listed as both sodium nitrate ($NaNO_3$), for acidic waste streams, and sodium hydroxide (NaOH), for caustic waste streams.

The WSB process simulation includes adiabatic and constant temperature mixing tanks, evaporators, and condensers. In the simulation, all tanks, evaporators, and condensers operate at a default pressure of one atmosphere, and all constant temperature mixing tanks and condensers operate at a default ambient temperature of 25 °C.

To expedite the simulation calculation, the solids in the OLI chemistry model are limited to precipitates known to form from the chemical species present in the feed solutions. These species include silver (Ag), which precipitates as Ag_2O, and the actinide species thorium (Th), uranium (U), neptunium (Np), plutonium (Pu), and americium (Am). Of the elemental species whose precipitation is suppressed, the alkali metals (+1 valence) and the alkali earth metals (+2 valence) are likely to be soluble at the concentrations present in the WSB process solutions, and the remaining metals and anions are present only at low concentrations and

---

® Aspen Plus is a trademark of Aspen Technologies, Inc., of Burlingame, MA. OLI is a trademark of OLI Systems, Inc., of Cedar Knolls, NJ. Excel is a trademark of Microsoft Corporation of Redmond, WA.

therefore would not contribute significantly to the overall solids content if they were to precipitate.

The simulation model is a steady state, continuous flow model. All output flows are reported in units of kg/year, and all cooling duties are reported in units of kcal/hr. These units assume that operation of the WSB processes continues at a constant rate 24 h/day, seven days/week, year-round.

OLI Chemistry Model Development

Aspen Technologies, Inc., utilizes an aqueous chemistry model developed by OLI Systems, Inc., in the form of thermodynamic equilibrium databases (called databooks in the OLI terminology). On the recommendation of OLI, the WSB evaporation and neutralization process model uses the most recently developed OLI databook, MSEPUB (Mixed Solvent Electrolyte, Public). Notably, the MSEPUB databook bases ionic speciation on the H_3O^+ ion, so acid concentrations are reported in terms of H_3O^+ concentrations. Historically, the OLI databooks have been used successfully to model nitric acid-based waste solution evaporators for the Hanford River Protection Project.

The OLI models for the ionic speciation and solubility of the actinide elements are of particular concern for the WSB model. The MSEPUB databook contains models for the most probable ionic states of all actinides of interest, including thorium(VI) (Th(IV)), uranium(VI) (U(VI)), neptunium(IV) (Np(IV)), plutonium(IV) (Pu(IV)), and americium(III) (Am(III)). Models for the majority of Th(IV), Np(IV), and Am(III) species are based on data from the Lawrence Berkeley Laboratory thermodynamic database.[1] Data for Pu(IV) were obtained from work at Hanford[2-6] and in Russia.[7] MSEPUB includes models for monatomic U(VI) ions[8] and for polyatomic U(VI) ions.[9] These models are restricted to cationic species, which predominate in acidic solutions, and MSEPUB contains only one solid U species, uranyl hydroxide ($UO_2(OH)_2$).

The MSEPUB model for U is not adequate for basic solutions, for which the precipitate takes the form of a sodium polyuranate ($Na_6U_7O_{24}.nH_2O$, where n is 12 or 16),[10] or, when fully dehydrated, crystalline sodium diuranate ($Na_2U_2O_7$).[11] This stoichiometry implies that, when an acid solution is neutralized by caustic, each U atom is associated with an ionic or molecular species such as $UO_2(OH)_3^-$ with a charge of -1 rather than a species with a neutral charge, as would be the case for a $UO_2(OH)_2$ precipitate. The predominant U species in acidic solution is UO_2^{2+}. Thus, substitution of $UO_2(OH)_2$ for $Na_2U_2O_7$ as the precipitate in caustic solution underestimates by 50% the amount of caustic that would have to be added to precipitate soluble U.

To address this discrepancy in the MSEPUB databook, a so-called private, or user, databook, named DRNA, was created with three new species added, a singly charged anion, $UO_2(OH)_3^-$, and soluble and solid forms of a sodium diuranate hydrate precipitate, $Na_2U_2O_7.5H_2O$. (The number of waters of hydration was selected to most closely match the stoichiometry of the most highly hydrated aqueous precipitate, $Na_6U_7O_{24}.16H_2O$.) The anion species was added so that the calculated solubility would drop as the solution becomes more caustic. For the WSB model, a simple empirical approach is taken in which the anionic uranyl species is assigned the simplest possible formula, $UO_2(OH)_3^-$; this approach is consistent with the uranyl species already present in the MSEPUB databook.

The private OLI databook DRNA was generated using the OLI ESP (Environmental Simulation Package) software. The $UO_2(OH)_3^-$ anion was added using the COMPLEX function, with UO_2OH^+ and OH^- specified as the basis ions. Initially, two species, UO_2OHOH

(The default name assigned by COMPLEX automatically and fortuitously differentiated this neutral species from the $UO_2(OH)_2$ species already in the public MSEPUB databook.) and $UO_2(OH)_2OH^-$ were added. Values for the thermodynamic state variables for $UO_2(OH)_2OH^-$ (the free energy (ΔG_{ref}) and the enthalpy (ΔH_{ref})) were then copied and transferred to a new species, $UO_2(OH)_3^-$. The two species created by the COMPLEX function then were deleted.

The last step in the creation of the database entry for $UO_2(OH)_3^-$ was to specify the logarithm of the equilibrium constant for the ionic reaction

$$UO_2(OH)_3^- \Leftrightarrow UO_2(OH)_2 + OH^-$$

No equilibrium data was available for hydroxide-complexed uranyl ions, so the equilibrium constant was empirically adjusted based on the "best value" equilibrium constants for chemically similar uranyl fluoride solutions.[12] By trial and error extrapolation that compared equilibrium constants for uranium oxyfluorides with those for uranium hydroxides, the value of the equilibrium constant for $UO_2(OH)_3^-$ was set at -4.8.

For sodium diuranate, two databook entries were made, one for a soluble aqueous species and one for a solid precipitate. The diuranate hydrate $Na_2U_2O_7.5H_2O$ was converted to an equivalent monouranate species, $NaUH_5O_6$, to simplify the chemistry within the OLI databook. Both the aqueous species and the precipitate were assigned the same thermodynamic properties.

Initial estimates for the free energy, enthalpy, and entropy (S_{ref}) were based on weighted averages of the thermodynamic properties of sodium diuranate ($Na_2U_2O_7$)[13,14] and water.[15] The enthalpy was adjusted by trial and error to match the calorimetrically-measured heat of solution of 44.03±0.27 kcal/mole for the diuranate species dissolving in 6 M HNO_3 at 25 °C.[16,17] Subsequently, the free energy and entropy were scaled by multiplying by the ratio of the adjusted enthalpy to the original estimate of the enthalpy. The molar volume was calculated from a sodium diuranate particle density of 3.93 g/cm^3, measured by the falling drop method.[18]

The entry for sodium diuranate also requires values for the logarithms of the equilibrium constants for the dissolution of the precipitate and the dissociation of the neutral aqueous species, by the reactions:

$$NaUH_5O_6(ppt) \Leftrightarrow Na^+ + UO_2(OH)_3^- + H_2O$$

and $$NaUH_5O_6(aq) \Leftrightarrow Na^+ + UO_2(OH)_3^- + H_2O$$

The equilibrium constants were adjusted to fit sodium diuranate solubility data for caustic waste supernate solutions with various concentrations of added NaOH.[11] By trial and error, the logarithm of the equilibrium constant for the aqueous species was set at 0.7, and the logarithm of the equilibrium constant for the precipitate was set at -3.9. Figure 1 compares the model predictions with the measured solubilities.

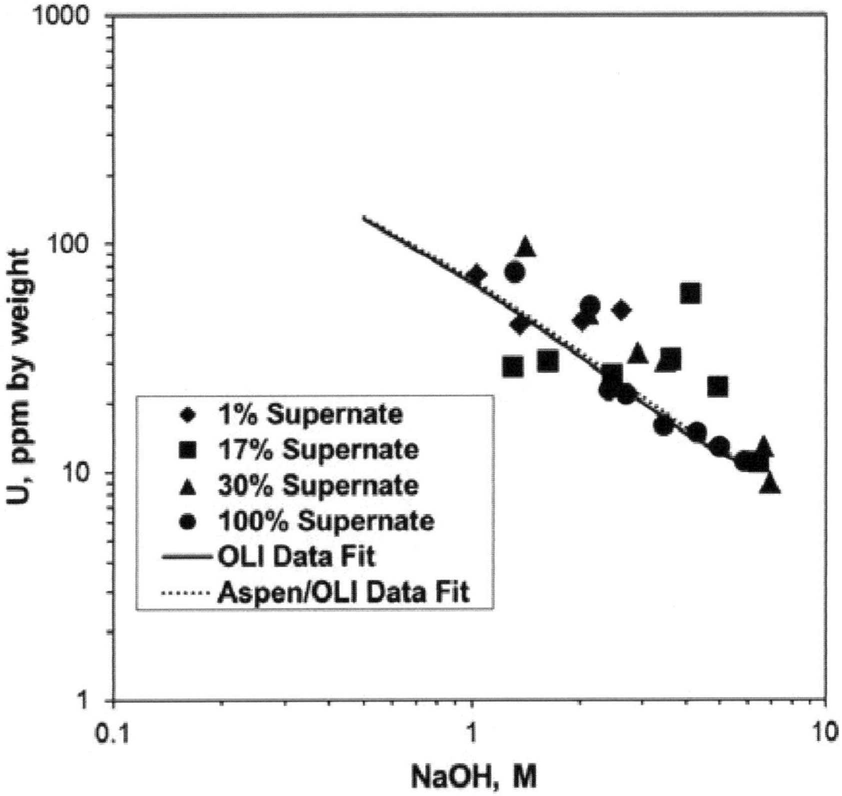

Figure 1. Comparison of OLI Model of Sodium Diuranate with Measurements of Hobbs and Karraker for Waste Tank Supernate/NaOH Mixtures

The sodium diuranate solubility correlation was tested by applying it to reproduce laboratory titrations of concentrated simulated transuranic waste solutions. Two titrations were reproduced, one starting at 30 °C and a second starting at 70 °C. At the start of the titrations, the solutions contained 0.75 L, with 290 g/L U, 0.34 M Al, 0.19 M Si, and 1.5 M HNO_3. Concentrated caustic (50 wt % NaOH) was added at a rate of 11.4 mL/min. Figure 2 compares calculated and measured pH values at 70 °C. According to the OLI calculations, the plateau in this figure at low pH is caused by neutralization of Al^{3+} and UO_2^{2+} and the precipitation of gibbsite ($Al(OH)_3$), boehmite (AlOOH), and uranyl hydroxide ($UO_2(OH)_2$). The gradual increase at intermediate pH's is due to the conversion of $UO_2(OH)_2$ to sodium uranates of varying stoichiometry, one of which, $Na_2(UO_2)_6(OH)_{14}.4H_2O$, was detected by x-ray diffraction analyses of the precipitates. The OLI model does not include any intermediate species between $UO_2(OH)_2$ and $Na_2U_2O_7.5H_2O$, so the model predicts an abrupt shift in the titration curve slightly into the caustic range. However, the OLI model accurately predicts the stoichiometry for the complete conversion to the diuranate species, as indicated by the agreement between the measured and predicted pH's where the pH levels out in the caustic region.

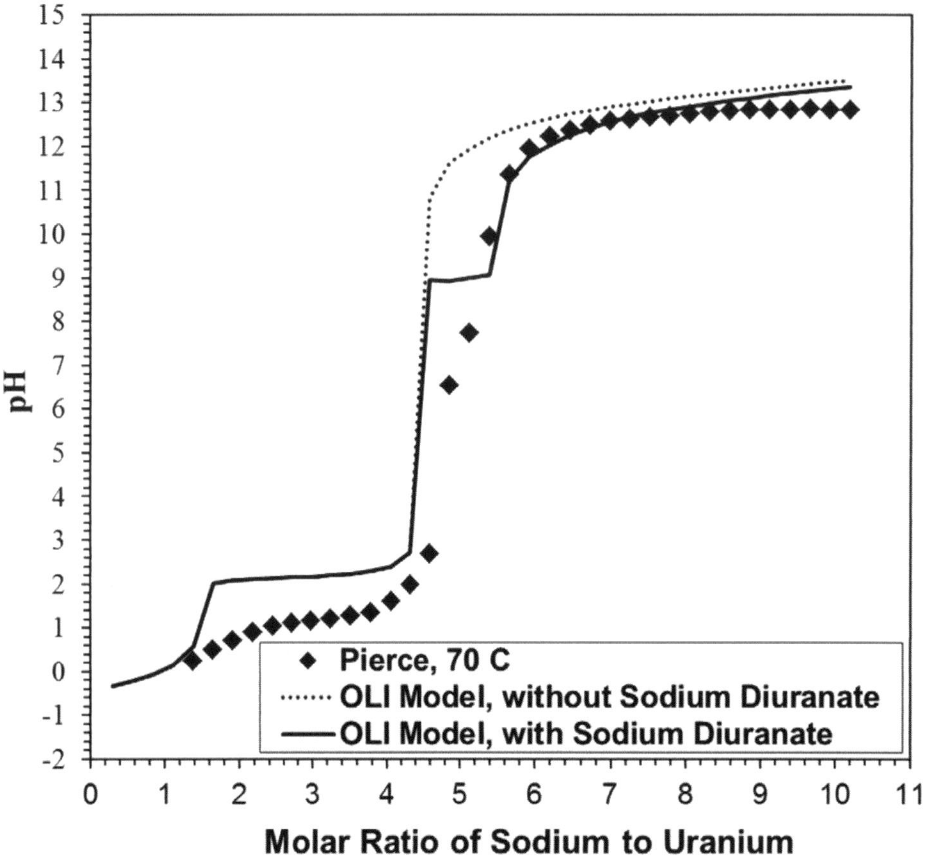

Figure 2. Comparison of OLI Uranium Solubility Predictions with Measurements of Pierce in a Solution with an Initial Temperature of 70 °C

A second private databank was added to correct density calculations for aqueous NaOH solutions. These density calculations are required for the HAW and LAW caustic neutralization streams. For a 50 wt % NaOH solution at 20 °C, the MSEPUB databook gives a density of 1.7176 kg/L, compared to a tabulated density of 1.5253 kg/L.[19] OLI Systems, Inc., was apprised of this error and provided a private databook called NAOH that contains corrected density calculations.

The solubility of NaOH must be modeled to ensure that the HAW and LAW caustic neutralization streams will not precipitate solids during their make-up and transfer to the neutralization tanks. At 25 °C, the solubility of NaOH in water is about 53 wt %. The solubility rapidly drops, however, as the caustic solution absorbed carbon dioxide (CO_2) from the air to form Na_2CO_3. Figure 3 depicts the variation of both measured and predicted Na_2CO_3 solubilities at 25 °C as a function of the NaOH concentration. It may be seen that Na_2CO_3 solubilities remain low at NaOH concentrations in excess of approximately 30 wt %, and then rise as the NaOH concentration decreases.

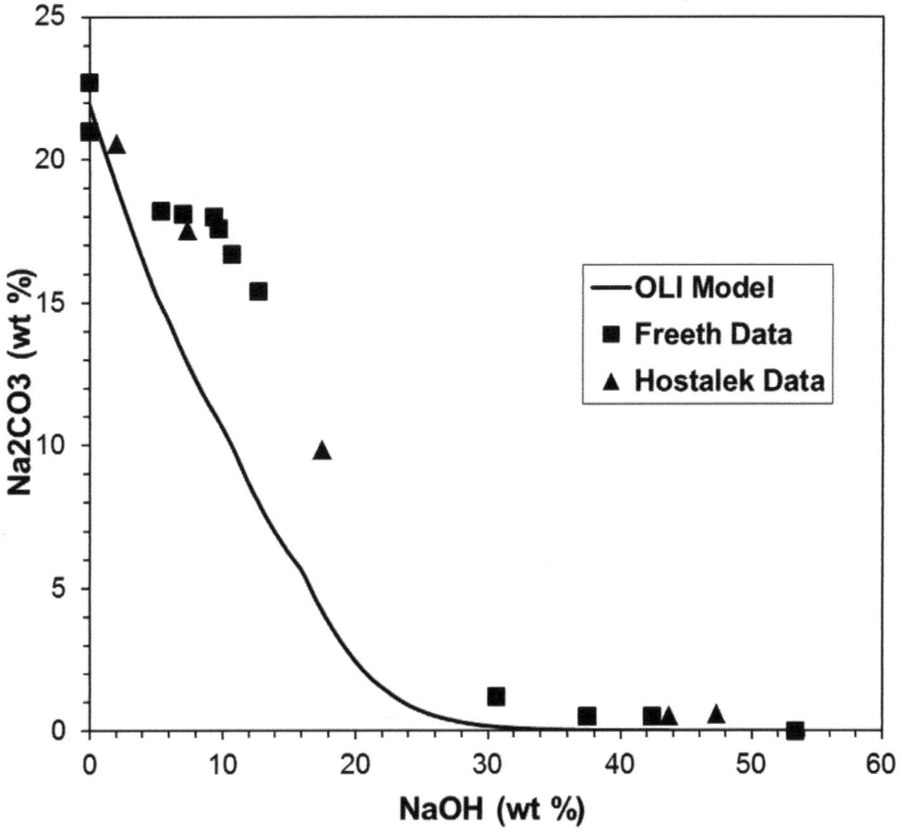

Figure 3. Comparison of OLI Carbonate Solubility Predictions with Measurements of Freeth[20] and Hostalek[21] at 25 °C

Simulation Methodology

In the Aspen Plus simulation package, chemical processing blocks are connected by flow streams. Using built-in vapor/liquid/solid equilibrium databases, Aspen iteratively calculates chemical and thermodynamic equilibria for each block in sequence. Figure 4 depicts the Aspen/OLI simulation of the WSB waste evaporation and neutralization process. Low activity waste (LAW) and high activity waste (HAW) are segregated into separate processing lines. The only connection between the HAW and the LAW processes is the transfer of the HAW evaporator condensate to the LAW feed tank. In the simulation, each processing line consists of a feed tank, an evaporator, and a neutralization tank. (The actual process design includes multiple tanks, pumps, and transfer lines which are not explicitly modeled by the simulation.) Each feed and neutralization tank is modeled as an adiabatic mixer to calculate the maximum possible temperature due to the heats of mixing and neutralization, followed by a heat exchanger to calculate the cooling duty required to bring the mixed stream back to ambient temperature. Each evaporator also includes a condenser, with the outlet stream set to ambient temperature and pressure.

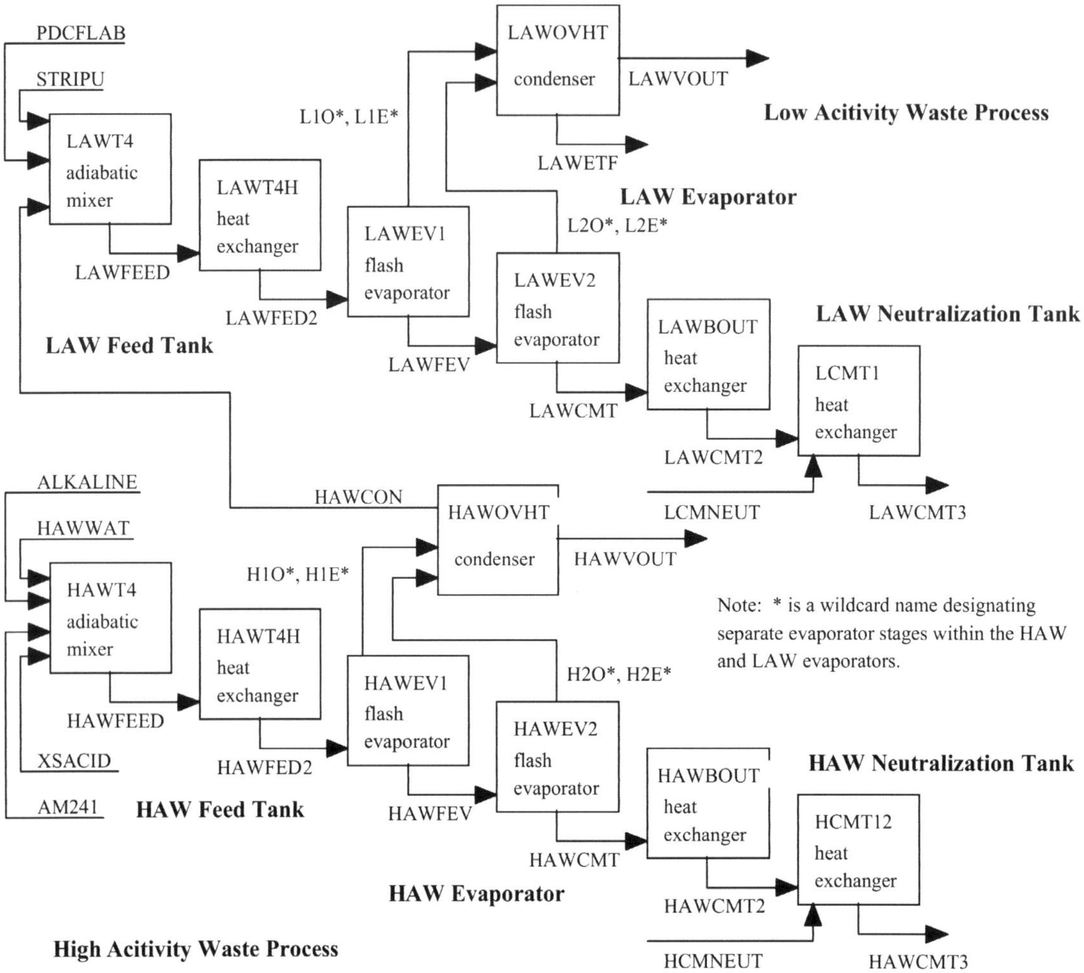

Figure 4. Schematic of Aspen/OLI Model of the Waste Solidification Building Evaporation and Neutralization Process

Feed-forward controls are used to set the flow rates of the HAW water dilution stream, the HAW and LAW caustic neutralization streams, and the fractions of flow that are evaporated in the HAW and LAW evaporators. The HAW water dilution stream is set to give an acid concentration of 1 M in the HAW evaporator feed stream, based on the combined total HNO_3 and NaOH concentrations in the other HAW feed streams. The HAW evaporator operation is set to concentrate the HAW evaporator bottoms acid concentration to 6 M, again based on the HNO_3 concentration. The LAW evaporator operation is set to concentrate the LAW evaporator bottoms to either 0.4 g/L Cl or 200 g/L U, whichever is reached first. The Cl limit is set to minimize corrosion in the process vessels and pipes, and the U limit is set to prevent U precipitation and solids handling problems in the LAW evaporator and piping downstream from the evaporator. The HAW and LAW caustic neutralization stream flow rates are set to add OH^- to the neutralization feed streams in concentrations that are approximately 10% in excess of those required to neutralize major acid and salt constituents, include HNO_3, H_2SO_4, and uranyl salts.

The target acid concentration in the HAW bottoms stream and the target Cl and U concentrations in the LAW evaporator bottoms stream are applied at the bottoms stream temperature. It is recommended that these targets be set lower than the desired concentrations in the cooled streams to account for solution density changes with temperature.

For the simulation calculations in this study, the target acid concentration in the HAW bottoms stream is set at 5.7 M HNO_3, and the target concentrations in the LAW bottoms stream are set at 0.37 g/L Cl and 185 g/L U. Caustic is added to the HAW and LAW neutralization tanks in amounts 10% in excess of those required to neutralize all principal anionic species. The composition of the caustic neutralization streams is set at 69 wt % H_2O, 30 wt % NaOH, and 1 wt % $NaHCO_3$, which is within the measured soluble range for caustic-carbonate solutions at 25 °C (see Figure 3).

Both the HAW and the LAW evaporators are modeled as 12-stage flash evaporators. Entrainment from each evaporator stage is modeled by splitting the bottoms stream from each stage so that a small fraction of the liquid, nominally set at 1 ppm by weight for each stage of both the HAW and LAW evaporators, is entrained with the vapor. The effective entrainment from all the stages of the HAW and LAW evaporators can be obtained by adding the entrainments from each of the stages; this gives an effective entrainment of about 10 ppm for each evaporator.

The staging of the evaporation process minimizes the fraction of vapor that is lost in any given stage and therefore minimizes the effect of partial vapor pressures on the vapor-liquid equilibria. In this way the modeling accounts for the effect of the process ventilation system in sweeping away vapors, without including any specific design component of the ventilation system. The only elements lacking in the model are the requirement for sensible cooling of the process ventilation gas, which should be small compared to cooling required for condensation and cooling of the vapor, and loss of volatile vapors due to evaporation in the condensers.

Discussion of Results

For the four simulation cases that were requested, tabulated comparisons of maximum temperatures, heat duties, and stream compositions at key locations in the WSB process follow. Table 2 compares heat duties (actually, cooling duties) for the evaporator feed tanks, evaporator condensers and bottoms tanks, and neutralization tanks. The highest cooling requirements are for the HAW and LAW condensers. The cooling requirements for the HAW and LAW evaporator bottoms streams are determined by the concentration maximums specified for these streams, and the cooling requirements for the HAW and LAW neutralization tanks are fixed by the salt concentrations in the HAW and LAW evaporator bottoms streams.

Table 2. Comparison of Heat Duties for Evaporator Feed Tanks, Evaporators, and Cementation Feed Tanks

	PDCF Feed Exp. Flow (kcal/h)	**PDCF Feed Max. Flow (kcal/h)**	**AFS Feed Exp. Flow (kcal/h)**	**AFS Feed Max. Flow (kcal/h)**
HAW Evaporator Feed Tank	108	114	42	45
LAW Evaporator Feed Tank	0	0	0	0
HAW Evaporator Condenser	17433	18420	7755	8468
LAW Evaporator Condenser	28772	30046	18519	19464
HAW Condenser Bottoms	342	361	150	164
LAW Condenser Bottoms	188	407	199	414
HAW Neutralization Tank	523	553	231	252
LAW Neutralization Tank	144	154	112	118

Results show that the LAW condensate stream carries most of the bulk effluent from the simulation process. The neutralized feeds to the HAW and LAW cementation processes both include precipitated solids. The HAW solids consist primarily of silver oxide (Ag_2O), and the LAW solids are comprised primarily of sodium diuranate ($Na_2U_2O_7.5H_2O$) and, for the AFS feed at the maximum flow rate and concentration, some uranyl hydroxide ($UO_2(OH)_2$).

Conclusion

An Aspen Plus/OLI model of the WSB process has been constructed and tested. The model includes separate mixing tanks, evaporators, and neutralization tanks for High Activity Waste (HAW) and Low Activity Waste (LAW) flow streams. The HAW and LAW evaporators separate the feeds into bottoms streams that are neutralized with concentrated caustic and overhead streams that are sent to condensers. The neutralized bottoms streams feed cementation processes, which are not modeled. The HAW condensate is mixed with the feed to the LAW evaporator, and the LAW condensate stream goes to an Effluent Treatment Facility (ETF), which is not part of the model. The Aspen Plus/OLI model has been tested successfully using WSB input data. Tabulated results are presented for four simulation cases, two each for feeds from the Pit Disassembly and Conversion Facility (PDCF) and from an Alternate Feed Stock (AFS).

As part of the modeling effort, a so-called private, or user, databook was created within OLI to model the neutralization and precipitation of uranium salts. The private databook extends the applicability of the existing OLI model into the basic pH range, where uranium forms sodium diuranate salts. The change in OLI was needed to accurately model the neutralization stoichiometry of uranium, which is present in high concentrations in the LAW evaporator bottoms.

References

1. Phillips, S. L., F. V. Hale, L. F. Silvester, and M. D. Siegel (1988), "Thermodynamic Tables for Nuclear Waste Isolation. Aqueous Solution Database," Lawrence Berkley Laboratory, NUREG/CR-4864.
2. Delegard, C. H. (1985), "Solubility of $PuO_2{\cdot}xH_2O$ in Alkaline Hanford High-Level Waste Solution," RHO-RE-SA-75 P, Rockwell International, Rockwell Hanford Operations, Richland, Washington.
3. Delegard, C. H.(1987), "Solubility of $PuO_2{\cdot}xH_2O$ in alkaline Hanford high-level waste solution," *Radiochim. Acta*, 41, 11-21.
4. Delegard, C. H. (2006), "Literature Review of Plutonium(IV) Hydrous Oxide Solubility in Dilute Sodium Hydroxide and in Dilute Nitric Acid Solutions," WTP-RPT-149, Rev. 0, Battelle – Pacific Northwest Division, Richland, Washington.
5. Krot, N., V. Shilov, A. Bessonov, N. Budantseva, I. Charushnikova, V. Perminov, and L. Astafurova (1996), "Investigation on the Coprecipitation of Transuranium Elements from Alkaline Solutions by the Method of Appearing Reagents," WHC-EP-0898, Westinghouse Hanford Company, Richland WA.
6. Peretrukhin, V. F., S. V. Kryutchkov, V. I. Silin, and I. G. Tananaev (1996), "Determination of the Solubility of Np(IV)-(V), Pu(III)-(VI), Am(III)-(VI), and Tc(IV),(V) Hydroxo Compounds in 0.5 - 14 M NaOH Solutions," WHC-EP-0897, Westinghouse Hanford Company, Richland WA.

7. Shilov, V. P., and A. M. Fedoseev (2003), "Solubility of Pu(IV) in weakly alkaline solutions (pH 9-14) containing silicate anions," *Radiochemistry (English transl.)*, 45 (5), 491-494; *Radiokhimiya*, 45 (5), 441-444.
8. Glushko, V. P., et al.(1978), Thermal Constants of Compounds, Academy of Sciences USSR, Vol. 8, Parts 1,2.
9. Grenthe, I., R. J. Lemire, A. B. Muller, C. Nguyen-Trung, and H. Wanner (1989), Chemical Thermodynamics of Uranium, NEA-TDB.
10. Wamser, C. A., J. Belle, E. Bernsohn, and B. Williamson (1952), "The Constitution of the Uranates of Sodium," *J. Am. Chem. Soc.*, 74(4), 1020-1022.
11. Hobbs, D. T., and D. G. Karraker (1996), "Recent Results on the Solubility of Uranium and Plutonium in Savannah River Site Waste Supernate," *Nucl. Technol.*, 114, 318-324.
12. A. M. Bond, A. M., and G. T. Hefter (1980), "Critical Survey of Stability Constants and Related Thermodynamic Data of Fluoride Complexes in Aqueous Solutions," IUPAC Chemical Data Series, No. 27, Critical Evaluation of Equilibrium Constants in Solution, Part A: Stability Constants of Metal Complexes, Pergamon Press, Oxford, UK. (Summary of Best Values table)
13. Katz, J. J., G. T. Seaborg, and L. R. Morss (1986), The Chemistry of the Actinide Elements, 2nd ed., Chapman and Hall, London.
14. Cordfunke, E. H. P., R. P. Muis, W. Ouweltjes, H. E. Flotow, and P. A. G. O'Hare (1982), "The Thermodynamic Properties of Na_2UO_4, $Na_2U_2O_7$, and $NaUO_3$," *J. Chem. Thermodynamics*, 14, 313-322.
15. Lide, D. R., ed. (1994), CRC Handbook of Chemistry and Physics, 75th ed., CRC Press, Boca Raton, FL.
16. Cordfunke, E. H. P., and B. O. Loopstra (1971), "Sodium Uranates: Preparation and Thermochemical Properties," *J. Inorg. Nucl. Chem.*, 33, 2427-2436.
17. Cordfunke, E. H. P. (1964), "Heats of Formation of Some Hexavalent Uranium Compounds," *J. Phys. Chem.*, 68(11), 3353-3356.
18. Karraker, D. G. (1994), "Uranium Settling Rates in SRS Waste Supernate," Westinghouse Savannah River Company Report WSRC-TR-94-058, Aiken, SC.
19. Green, D. W., ed. (1984), Perry's Chemical Engineers' Handbook, 6th ed., McGraw-Hill, New York, NY, p. 3-83.
20. Freeth F. A. (1922), "System: Na_2O-CO_2-NaCl-H_2O, Considered as Two Four-Component Systems," *Philos. Trans. R. Soc. London, Ser. A*, 223A, 35-87.
21. Hostalek, Z. (1956), "The Ternary System Water-Sodium Carbonate-Sodium Hydroxide," *Chem. Listy*, 50, 716-720.

431719 Small Scale Testing of a Floating Ion Exchange Bed for Nuclear Waste Treatment

Abstract:

Technetium-99 is a waste isotope that is being consdered for separation in high level nuclear waste at Department of Energy sites like Hanford in Washington state and Savannah River in South Carolina. Technetium-99 has a half-life of 211,000 years and is highly mobile in groundwater. Ion exchange of the liquid wastes is one processing option to remove the technetium that is present in solution as pertechnetate.

SuperLig® 639 (IBC Advanced Technologies) resin was tested under challenging conditions in the current work. A nonradioactive surrogate waste solution containing perrhenate ion as the trace representative for pertechnetate was used, its sodium molarity being atypically high at 8 M (such wastes are typically 5-6 M). The higher molarity would allow more efficient throughput in a large process. The current work was done at the bench scale with a 10 mL resin bed. One goal of the work was to examine bed behavior when it floats during loading (feed specific gravity 1.34), and sinks when eluted with water at 65 °C. Feed upflow was found to float the bed upward as a single and stable unit. The work saw 50% breakthrough after more than 50 bed volumes at 3 bed volumes per hour. The resin bed was then smoothly sunk with a downflow of water. Elution at about 1 BV/hr produced an output peak of perrhenate concentration greater than 20 times the feed concentration.

The work is a first demonstration that the ion exchange process can be operated at high feed sodium molarity with SuperLig® 639 resin. Further work will be at larger scale. Savannah River National Laboratory has an ion exchange modeling capability to relate this performance to that of resin separating pertechnetate from actual waste.

434544 Characterization of the Next Generation Solvent Used to Remove Cesium at the Savannah River Site

Abstract:

A Next Generation Solvent (NGS) was developed through research funded by the Department of Energy (DOE) to improve cesium decontamination relative to the current Caustic-Side Solvent Extraction Solvent (CSSX) (decontamination factor from 12 to greater than 40,000) with an increased performance of high-level waste processing in the Modular Caustic-Side Solvent Extraction (CSSX) Unit (MCU) and the Salt Waste Processing Facility (SWPF) at Savannah River Site (SRS) . The majority of NGS development was conducted at Oak Ridge National Laboratory while testing was conducted at the Savannah River National Laboratory and deployment at the MCU. The changes in solvent chemistry (and corresponding mass transfer and solvent recovery changes) require new analytical methods to track the solvent compositional changes and corresponding solvent extraction performance. This talk evaluates new analytical methods to characterize the solvent and a correlation is established between solvent extraction performance and the chemistry changes in the solvent.

Innovative Idea to Accelerate Tank Closure in DOE Savannah River Site

Robert Chang, Pete Hill, Aaron Staub
Savannah River Remediation LLC
Aiken, SC
Robert.Chang@srs.gov; Peter.Hill@srs.gov; Aaron.staub@srs.gov

ABSTRACT

Treatment and disposition of salt waste is the critical path to completion of Savannah River Site (SRS) Liquid Waste (LW) Disposition Program. The Salt Waste Processing Facility (SWPF) is the integral part of the overall strategy to complete the SRS tank waste mission to process ~37 Mgal of radioactive salt waste. Schedule delay for the SWPF project has placed a significant challenge to meet tank closure commitments at SRS in Aiken, South Carolina.

Actinide Removal Process (ARP) and Modular Caustic Side Cesium Extraction Unit (MCU), interim salt processing facilities, were constructed in 2008 to bridge the crucial period before startup of SWPF. Current process is to transfer a batch of 3,715-gallon salt solution to a tank, strike the salt batch with MST to absorb actinides. The MST-stricken batch is then transferred to a tank to filter out solids. The filtrate, Clarified Salt Solution (CSS), is then fed to MCU to remove cesium. The filtering process is one of bottlenecks and it requires time consuming chemical cleaning steps once every 20 filtering batches. The need to increase ARP/MCU throughput to beyond design basis rate becomes more pressing. The current cycle time for qualifying a one million gallon batch of salt waste for Waste Acceptance Criteria (WAC) program, on average, is about 3 month and 1 week. To take advantage of the WAC idle time, the "Large Tank Strike" project is proposed. By adding MST strike to the 1 million gallon batch instead of 3,715 gallon mini-batch during the WAC cycle time period while waiting for WAC qualification lab result, a bulk of solids is expected to settle below the pump suction head of Tank 21. Based on the Particle Settling Rate Calculations (Henley, E.A., Yueng, M.R., 2014), more than 90% of solids are expected to settle out during the 3 months and 1 week WAC cycle time period. The upper CSS solution is transferred to a tank for filtering through existing primary and secondary filtration systems. With much lowered solid loadings, it is expected the downtime and filter cleaning time frequency will be significantly less. The project is critical for DOE to mitigate impacts to Federal Facility Agreement (FFA) tank closure dates resulting from SWPF construction delays.

ABOUT THE AUTHORS

Bob Chang has more than 20 years of hands-on process engineering and computer simulation modeling experience in nuclear and hazardous waste industries. He is a Fellow Engineer at Savannah River Remediation LLC (SRR). He has presented and published several technical papers in Journals and conferences and holds three patents. Mr. Chang holds a Certified Modeling and Simulation Professional (CMSP) from Modeling and Simulation Certification Committee (MSPCC). Over his 20 plus years tenure at SRS, Mr. Chang has made substantial contributions to many facilities and projects that are dependent on simulation modeling to cut cost and to improve process performance. He received a Bachelor Degree in Chemical Engineering from Tatung Institute of Technology, Taipei and a Master Degree in Chemical Engineering from Oklahoma State University.

Pete Hill has 26 years of experience in Nuclear High-Hazard Facility operations including Reactor Operations, Vitrification Operations, and nuclear waste storage & pretreatment. He currently is responsible for facility integration, process optimization, and strategic planning for SRR. Mr. Hill is a Certified Six Sigma Black Belt. He graduated from Carnegie Mellon University with a B. S. in Mechanical Engineering.

Aaron V. Staub, a Senior Engineer at SRR, graduated from University of Alabama with a B.S. in Chemical Engineering. He has fifteen years of experience in the immobilization of high level and low level wastes at SRS. He is currently the Engineering Manager for the Large Tank MST Strike Project and supports Tank Farm Engineering for SRR.

Innovative Idea to Accelerate Tank Closure in DOE Savannah River Site

Robert Chang, Pete Hill, Aaron Staub
Savannah River Remediation LLC
Aiken, SC
Robert.Chang@srs.gov; Peter.Hill@srs.gov; Aaron.staub@srs.gov

INTRODUCTION

Treatment and disposition of salt waste is the critical path to completion of Savannah River Site (SRS) Liquid Waste (LW) Disposition Program. During the period prior to startup of the Salt Waste Processing Facility (SWPF), salt waste disposition will continue through the Actinide Removal Process (ARP) and Modular Caustic Side Solvent Extraction (CSSX) Unit (MCU) facilities. Deliquification, Dissolution, and Adjustment (DDA) processing was required prior to startup of the ARP/MCU facilities to enable continued tank closure activities, to sustain sludge disposition activities in the Defense Waste Processing Facility (DWPF), and to minimize continued limited use of old-style tanks. During DDA phase, approximately 2.8 million gallons (Mgal) of dissolved salt solution from Tank 41 and associated adjustment streams were dispositioned. Another 4.1 Mgal of salt waste has been dispositioned through ARP/MCU since startup in April 2008.

There are ~35 Mgal of radioactive salt waste currently stored in new and old-style tanks (old-style tanks do not have full secondary containment). Some old-style tanks have known leak sites. Refer to Figure 1 for schematics of Liquid Waste Program in Savannah River Site. The program baseline recognizes challenges from the further delay of SWPF and its concomitant effect on regulatory milestones for DOE to meet future waste removal and tank closure commitments at Savannah River Site (SRS) in Aiken, South Carolina.

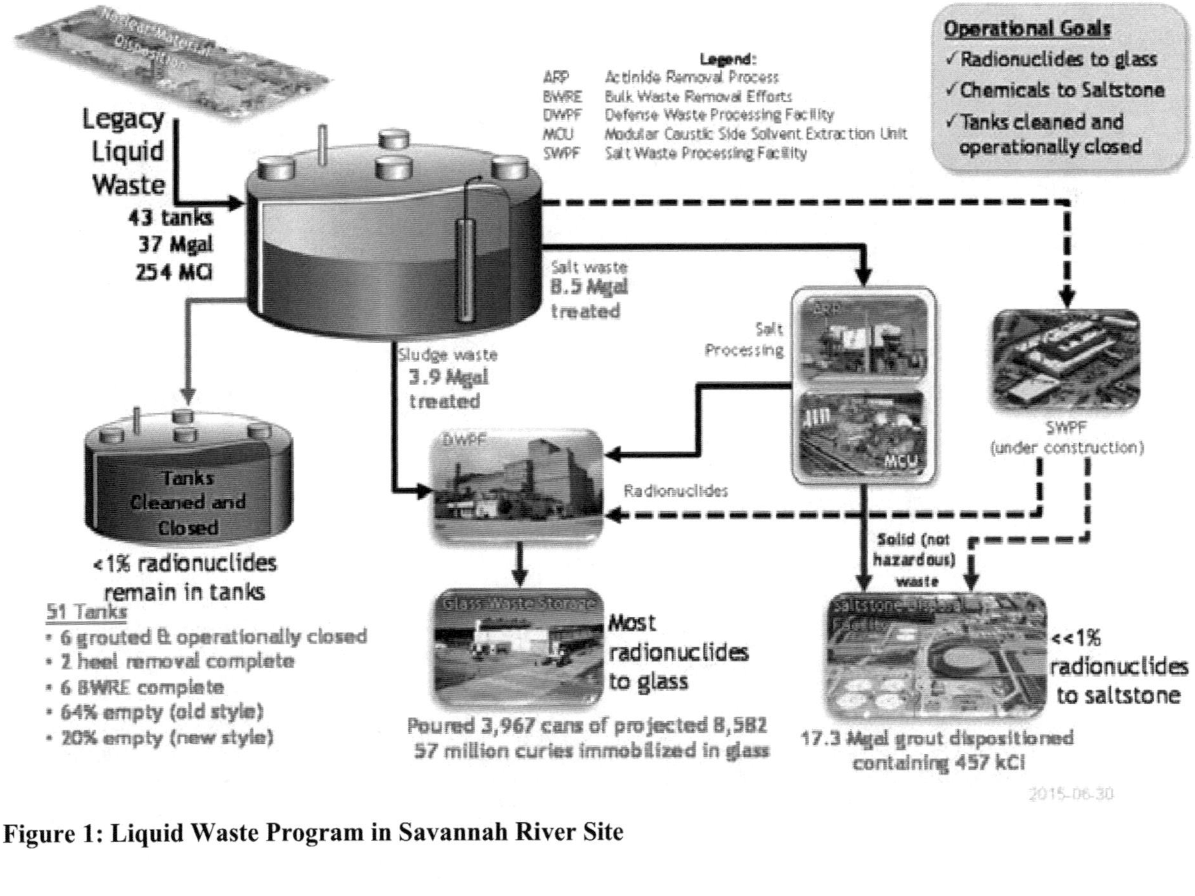

Figure 1: Liquid Waste Program in Savannah River Site

An interim salt processing facility, ARP and MCU, was constructed and started up in 2008 with a design life of three-years to bridge the crucial period before startup of the SWPF. With delay of SWPF, however, ARP/MCU has been enhanced and improved to provide a longer term option for salt disposition. The original goals of ARP/MCU process were (1) treat salt solution prior to start of SWPF and (2) provide operational experience and lessons learned for the SWPF project. The SWPF facility is designed with 7.0 Mgal/yr nominal salt processing rate. The design basis of ARP/MCU facility is 1 Mgal/yr. Refer to Figure 2 for the composite waste inventory in SRS. In FY2013, the ARP/MCU set a record of processing 1.3 Mgal of salt solution. The latest projected start date for SWPF is September 2018. The System Plan (Chew, D.P. & Hamm, B.A., 2014) mitigates the effect of these challenges by continuing to resequence the SRS LW program in a way that integrates the optimum combination of facility operating schedules, supporting infrastructure upgrades, and workforce management actions to enable the completion of as many deliverables as possible under reduced budget projections. The overarching objective of the System Plan is to ensure safe storage of the waste and minimize extension of the remaining time at risk associated with legacy high level waste storage in aging tanks.

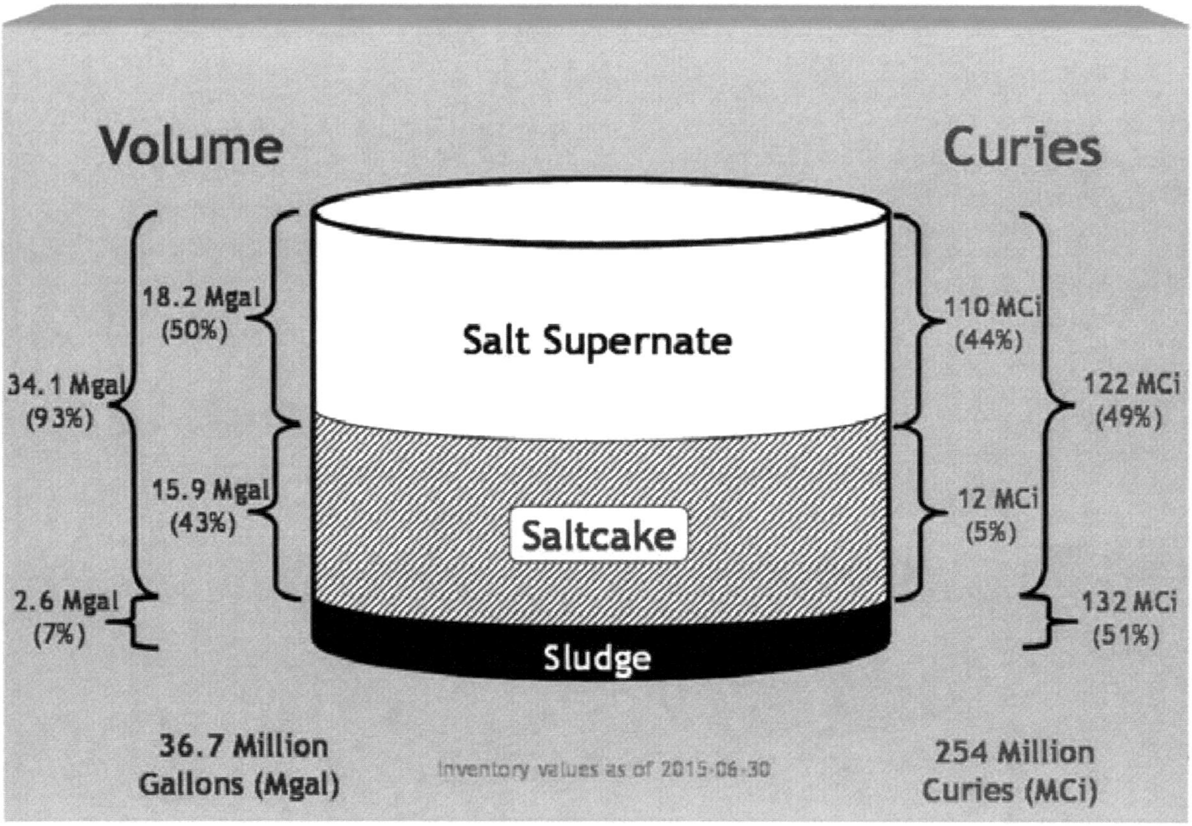

Figure 2: Composite Waste Inventory in Savannah River Site

FEDERAL FACILITY AGREEMENT (FFA)

The EPA, DOE, and SCDHEC executed the SRS FFA on January 15, 1993, which became effective August 16, 1993. It provides standards for secondary containment, requirements for responding to leaks, and provisions for the removal from service of leaking or unsuitable Liquid Waste (LW) storage tanks. Tanks scheduled for operational closure may continue to be used, but must adhere to the FFA schedule for operational closure and the applicable requirements contained in the Tank Farms' industrial wastewater treatment facility permit. An agreement between DOE, SCDHEC, and EPA (Statement of Resolution of Dispute Concerning Extension of Closure Dates for Savannah River Site High-Level Radioactive Waste Tanks 19 and 18 effective in November 2007) modified the FFA by providing for the submission of Waste Determination document for F- and H-Tank Farms and including end dates for bulk waste removal and the operational closure of each old style tank. The FFA requires SRS to complete operational closure the last old-style tank no later than 2022.

In order to accelerate tank closure in Savannah River Site, there is a need to increase ARP/MCU throughput to beyond design basis rate. The current cycle time for qualifying a one million gallon batch of salt waste for Waste Acceptance Criteria (WAC) program is about 3 month and 1 week. The "Large Tank Strike" project is to take the advantage of the extended WAC qualification time by adding MST to the 1 million gallon batch instead of a 3,715-gallon mini-batch and settling MST during the WAC cycle time period. It is expected that the bulk of solids would settle below the transfer pump suction head, remaining inside Tank 21 and not be transferred into Tank 49. The solids staying in Tank 21 will be eventually transferred to the Sludge Feed Tank, Tank 40, and to be fed as sludge waste to the Defense Waste Processing Facility (DWPF). Based on the Solid Particle Settling Rate calculations (Henley, E.A., Yueng M.R., 2014), it predicted 90%–95% of solids would be precipitated out during the WAC cycle time period. The upper Clarified Salt Solution (CSS) solution is then transferred to a tank to be filtered through the existing primary and secondary filtration systems. Since the solids loading in filters is much lowered in this Large Tank Strike case, it is expected the filtering system's downtime will be significantly lowered and it would reduce required chemical cleaning frequency from current once every 20 batches to once every 200 batches. The resulting throughput improvement for ARP/MCU arising from implementation of the Large Tank Strike concept will help mitigate delays in meeting regulatory milestone ensuing from delays in SWPF construction.

Actinide Removal Process (ARP)

Interim salt processing used for lower activity salt waste entails use of ARP and 512-S to remove actinides and of MCU to remove cesium before the resultant low-activity salt waste stream is sent to Saltstone Process Facility (SPF) for encapsulation in grout. To serve as a source tank of feed to ARP/MCU, a salt tank has to have a relatively low concentration of ^{137}Cs.

The ARP facilities provide the capability to treat salt solution from the Tank Farms to remove soluble radioactive strontium (Sr) and actinides by addition of monosodium titanate (MST). Salt feed for ARP will be prepared and qualified in a blend tank. The ARP operates with two small MST strike tanks. Tank 49 receives qualified feed material from the blend tank (Tank 21) to feed the process. The waste is received into two MST Strike Tanks in 241-96H. Waste received in MST Strike Tank #1 and Tank #2 is adjusted with water to approximately 5.6 Molar (M) sodium concentration to provide optimum conditions for sorption of Sr-90 and actinides onto the MST.

Following the addition of MST to either Strike tank, the contents are agitated for a reaction period between 4 and 12 hours based on the radioactivity of the soluble actinides to be removed. The resulting slurry is then transferred from either of the Strike Tanks into Late wash Precipitate Tank (LWPT) located in 512-S.

The batch received in the LWPT is circulated through a cross-flow filtration process yields a CSS which is received in the Late Wash Hold Tank (LWHT). The CSS is sent to MCU for Cs removal. The solids are concentrated in the LWPT over a number of batches depending on the entrained solids loading of the waste incoming to ARP. The concentrated solids heel in the LWPT is then washed and filtered to remove soluble sodium salts. The washed MST/Sludge slurry is transferred to DWPF for vitrification processing. Spent wash water from the solids washing operation is transferred to Tank 50 to feed to SPF. Refer to Salt Batch Plan (Le, T.A., 2014) for more details on Salt Batching Plan.

Modular Caustic Side Solvent Extraction Unit (MCU)

MCU utilizes the caustic side solvent extraction (CSSX) process to remove Cs from the prepared salt waste solution. The solvent extraction process uses an organic solvent to remove Cs ions from the waste stream. The solvent and salt solution are fed to a bank of annular centrifugal contractors which ensure mixing of the waste in the annular region and extraction of the solvent and Cs from the salt waste stream. The Cs is then stripped from the solvent using a dilute nitric acid stream in another bank of contactors. The Strip Effluent (SE) is sent to DWPF for vitrification. The solvent is continuously recycled during this process. The decontaminated salt solution (DSS) resulting from solvent extraction is sent to Tank 50 and then transferred to SPF for disposal.

There are four main stages in CSSX: (1) Extraction Stage, (2) Scrub Stage, and (3) Strip Stage, and (4) Wash Stage. <u>Extraction stage,</u> uses solvent consisting of 0.01 M BobCalix6 extractant, 0.2 M Cs-3 modifier, and Isopar®L diluent to contact with alkaline waste stream in a series of seven countercurrent centrifugal contactors. <u>Scrub Stage</u>, following cesium extraction; the solvent is scrubbed with 0.05 M nitric acid in a series of two centrifugal pumps to remove other soluble salts from the solvent stream to aqueous phase. Strip stage, in a series of seven countercurrent centrifugal contactors, uses solvent contacted with a 0.001 M nitric acid stream containing a small quantity of cold cesium to transfer the cesium to the acid stream. Wash Stage, in which the used solvent is then washed using 0.01 M caustic in a series of two centrifugal pumps and recycled.

CURRENT ARP/MCU PRACTICE AND FLOWSHEET

ARP uses a primary filter and a secondary filter to separate solids formed by actinides absorbed by Monosodium Titanate (MST). The current process is to transfer a batch of 3,715 gallons of salt solution to a tank, then, strike the salt batch with MST for 8 hours to absorb actinides. Afterward, the MST treated batch is transferred to a tank to filter out solids using cross flow filtration. The filtrate, Clarified Salt Solution (CSS), from the filtration process, is then fed to MCU to remove cesium. Two product streams result from the MCU process: cesium Stripped Effluent (SE) and Decontaminated Salt Solution (DSS). Refer to Figure 3 for the existing ARP/MCU Salt Processing Flowsheet.

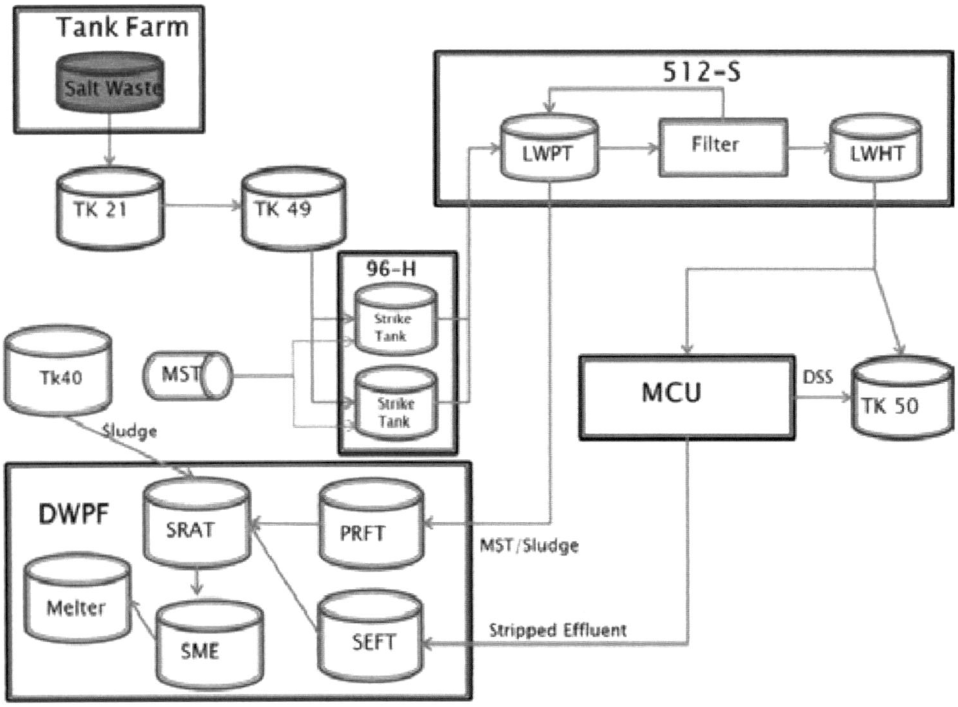

Figure 3: Existing Salt Processing Flowsheet in SRS (ARP/MCU flowsheet)

The filtering process is the bottleneck and it requires frequent chemical cleaning, very time consuming steps, once every 20 batches of filtration.

As a result of the combination of delay in the SWPF project and potential shortfall in out-years' congressional funding, the FAA committed closure dates for old-style tanks are at risk. The need to increase ARP/MCU throughput to beyond design basis rate is becoming more pressing.

PROPOSED INNOVATIVE LARGE TANK STRIKE FLOWSHEET

One of main objectives of the Large Tank Strike (LTS) project is to increase the salt processing capability at SRS. The primary unit operation in LTS involves the addition of Monosodium Titanate (MST) in a million gallon tank (Tank 21) which will sorb and separate soluble actinides and strontium prior to additional waste treatment. The actinide-laden MST will settle at the bottom of Tank 21 while the CSS is transferred to Tank 49 for further waste treatment. The LTS process intends to improve filtration throughput at 512-S by retaining settled MST solids within

Tank 21, while it is assumed the bulk of MST solids (≥90% by weight) will not be transferred forwards to Tank 49. Refer to Figure 4 for the LTS flowsheet.

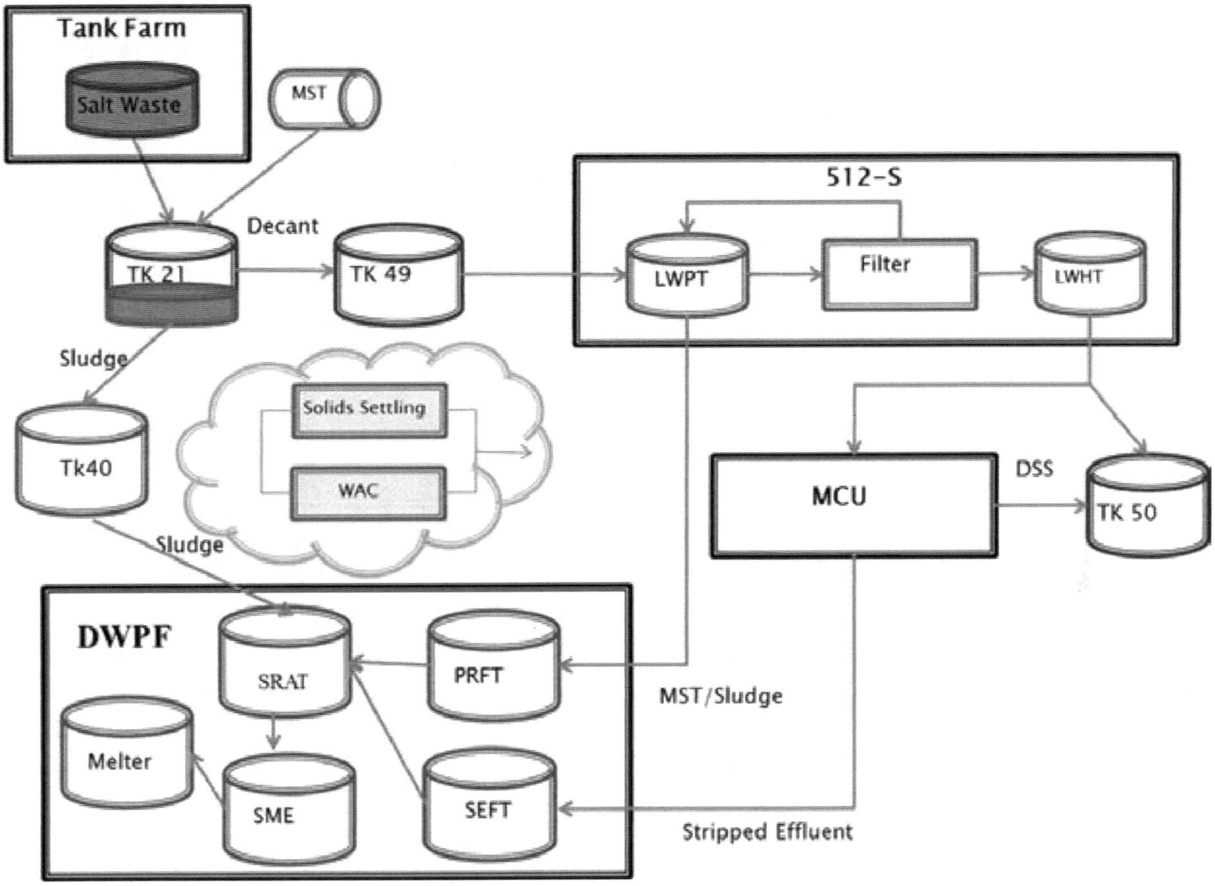

Figure 4: Proposed Innovative Large Tank Strike (LTS) Flowsheet

To validate the assumption mentioned above, a calculation for the solid particles settling rate was developed using the assumed physical property data for MST and salt solution as well as inputs for Tank 21 geometry and transfer pump elevation to establish required settling times to ensure that a given mass fraction of the total MST has settled below the pump suction. Multiple tank fill levels and settling fractions (e.g., 90% MST settled, 95% MST settled) are evaluated. Particle size (Staub, A.V., 2014) and other inputs parameters (Ledbetter, J., 2011) are critical parameters in calculating terminal velocities. Table 1 gives an MST particle size distribution. Each percentile value represents the cumulative mass fraction with respect to the listed diameter. For example, 10% of the mass resides in particle of 2.503 μm diameter and less. The 90% settled condition analyzed means only particles with diameter at the 10th percentile level and less (i.e., ≤ 2.503 μm) remains unsettled. MST particle density is 3.068 g/ml and the minimum density of sludge solids (from DWPF recycle or from insoluble solids in salt solution) is 2 g/ml.

Table 1. MST Particle Size Distribution

Percentile	Diameters (μm=10^{-6}m)
5	1.9
10	2.503
25	3.29
40	3.92
50	4.34
70	5.32
90	7.14
95	8.30

ANALYTICAL METHODS AND CALCULATIONS

The following gives the methodology for determining particles' settling terminal velocities and settling times.

Stokes' Law is applicable to solutions with low particle concentrations; such that each particle behaves as if were only in the presence of liquid. If the concentrations of particles are high, particles will interact with each other and Stoke's Law will not be applicable. Stokes' Law is applicable for solids' concentration up to about 7.5 weight percent (Lee, E.D, 1996). With given limitation, it is determined that Stokes' Law is applicable to the LTS cases as solids' concentrations is less than 7.5 wt% as shown in calculations performed by E.A. Henley, M.R. Yeung (2014).

Calculation of Settling and Terminal Velocities via Stoke's Law

Each particle has three forces acting on it: (1) the downward force of gravity (weight), (2) the upward force of buoyancy, and (3) the upward force of drag. Agglomeration and the effects of molecular collisions on very small particles are conservatively ignored. If particle agglomerate, they will settle more quickly. The effect of these collisions is to lower the drag force and allow particles to settle more quickly. Considering molecular collisions would decrease the settling time slightly. So, it is conservative not to consider agglomeration (Baron, P.A., Willeke, K., 2001) and the effects of molecular collision on very small particles (De Nevers, N., 1995).

$$\sum F = ma = F_g - F_b - F_d = \rho_p V g - \rho_f V g - F_d = \rho_p \left(\frac{\pi}{6}\right) D^3 g - \rho_f \left(\frac{\pi}{6}\right) D^3 g - F_d \qquad \text{(Eq. 1)}$$

where a = particle acceleration [m/s^2]=du/dt,

m = MST particle mass [kg],

F_g = particles weight [N],

F_b = buoyant force [N],

F_d = drag force [N],

V = MST particle volume (based on sphere) $[m^3] = \left(\frac{\pi}{6}\right) D^3$,

ρ_p = MST particle density [kg/m^3],

ρ_f = fluid density [kg/m^3],

g = acceleration due to gravity [m/s^2],

D = MST particle diameter [m], which is given by the 10th percentile particle size (Table 1) for the time for 90% of the particles to settle, or the 5th percentile size for the time to 95% of the particles to settle.

The drag force as a function of velocity was found by Stoke's to be:

$$F_d = 3\pi\mu D u \qquad \text{(Eq. 2)}$$

where μ = dynamic viscosity of the fluid [Pa-s],

u = particle velocity [m/s], which is a function of time, t[s].

The terminal velocity can be found by setting ΣF equal to zero:

$$u_t = \frac{gD^2}{18\mu}(\rho_p - \rho_f) \qquad \text{(Eq. 3)}$$

This is true as long as the Reynolds number associated with relative flow of fluid around the particle is less than or equal to 0.3.

$$Re = \rho_f u_t D/\mu \qquad \text{(Eq. 4)}$$

From equations 3 and 4 and using Inputs 2, 7, & 3 for ρ_f, ρ_p, and μ (Henley, E.A. & Yeung M.R., 2014),we can calculate the maximum particle diameter for which Stoke's Law is applicable by setting Re=0.3:

$$D = \sqrt[3]{\frac{Re18\mu^2}{g\rho_f(\rho_p - \rho_f)}} = 127.6[\mu m] \qquad \text{(Eq. 5)}$$

Because the particle sizes of interest are less than 10 μm (Input 11, Henley, E.A. & Yeung M.R., 2014), Stoke's Law is does not need a correction for larger sized particles. If calculations were needed for larger particles (Re>0.3), methodology from Section 8.2.2 of the paper by De Nevers, N. (1995) could be used.

The transient particle settling velocity is calculated from equations 3 and 2:

$$ma = \rho_p\left(\frac{\pi}{6}\right)D^3\frac{du}{dt} = \rho_p\left(\frac{\pi}{6}\right)D^3 g - \rho_f\left(\frac{\pi}{6}\right)D^3 g - 3\pi\mu D u, \text{ or} \quad \text{(Eq. 6)}$$

$$\frac{du}{dt} + \frac{18\mu u}{\rho_p D^2} = g\left(1 - \frac{\rho_f}{\rho_p}\right) \quad \text{(Eq. 7)}$$

This is an ordinary differential equation of the form $\frac{du}{dt} + Au = B$ with a solution of $u = B/A(1 - a^{-At})$, for u=0 at t=0, or

$$u = \frac{g\left(1-\frac{\rho_f}{\rho_p}\right)}{\frac{18\mu}{\rho_p D^2}}\left(1 - e^{-\frac{18\mu t}{\rho_p D^2}}\right) = u_t\left(1 - e^{-\frac{18\mu t}{\rho_p D^2}}\right) \quad \text{(Eq. 8)}$$

Calculation of Settling Time

The falling height, which is the difference between the fill height and the pump height, is divided by the terminal velocity to give the settling time.

$$t_s = \frac{h_{fall}}{u_t} = \frac{h_{fill}[in] - h_{pump}[in]}{39.37\left[\frac{in}{m}\right] * u_t\left[\frac{m}{s}\right]} \quad \text{(Eq. 9)}$$

Settling and Terminal Velocities

Figure 5 shows the settling velocity as a function of time for a 2.5 μm particle with a particle density of 3.068 g/ml (Henley, E.A. & Yeung M.R., 2014). The terminal velocity is reached in about a microsecond, so only the terminal velocity is needed. Thus, for settling time calculations, which occurs over many days, it is appropriate to just use the terminal velocity.

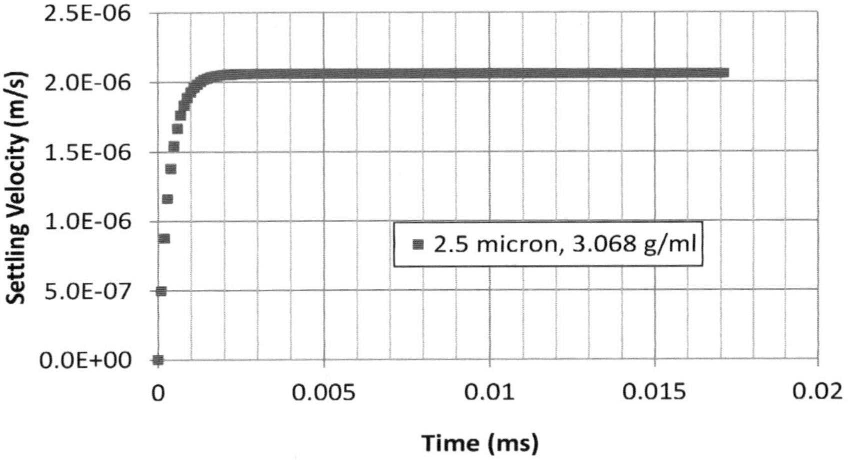

Figure 5: Settling velocity vs. time for a 2.5 μm particle, with a particle density 3.068 g/ml

The terminal velocities for the 90th (2.5μm) and 95th (1.9 μm) percentile particles with the density of 3.068 g/ml are 4.05 inches/day and 7 inches/day respectively. The terminal velocities for the 90th (2.5μm) and 95th (1.9 μm) percentile particles with the density of 2.5 g/ml are 2.75 inches/day and 4.77 inches/day respectively.

SRR Integrated LWO (Liquid Waste Operations) Model

An integrated model was developed to model & simulate SRS Liquid Waste operations since 2007. Modules (individual facility) as shown in Figures 1, 3, and 4 are developed piece by piece; and eventually integrated to a SRR Integrated LWO model. An independent "System Flow Model DWPF Validation Study" was developed by MITRE in 2013. The result predicted by the MITRE model matched predictions by the in-house SRR Integrated Model closely.

From 1/9/15 to 2/9/15, ARP/MCU processed a record 246,859 gal of waste which is equivalent to 2.91 Mgal/yr salt processing throughput rate. The SRR Integrated LWO model, loaded with same operating parameters of the record runs, has predicted the annual salt processing rate of 2.88 Mgal/yr which matches actual processing rate of 2.91 Mgal/yr with only 1% difference.

Conclusion from Settling Time Calculation

For LTS flowsheet, settling MST is to take place while the Tank 21 is performing Waste Acceptance Criteria (WAC) qualification, which would take on average of 3 months and 1 week (~98days). Therefore, as long as no less than 90% by weight of solids would settle down during the period of WAC qualification process, the LTS flowsheet would meet project objective. (Staub, A.V., 2014)

Based on Table 2, if particle density is assumed to be 3.068 g/ml, it would take no more than 47.5 days and 82.3 days respectively to settle 90% and 95% by weight solids inside Tank 21 to below 50".

Table 2: Settling times [days] for all cases

	Fill H (inch)	90% 2.5 g/ml	90% 3.068 g/ml	95% 2.5 g/ml	95% 3.068 g/ml
to fall to 50 inches	383.0	69.8	47.5	120.9	82.3
	372.8	67.7	46.1	117.2	79.8
	344.5	61.8	42.0	106.9	72.8
	316.3	55.8	38.0	96.7	65.8

For the existing ARP/MCU flowsheet, one of most restrictive bottlenecks is the 512-S filtration process, caused by low filtration rate (5 gpm on average) and frequent filter washing/cleaning required (36 hours on average for each filter cleaning/washing). Projected benefits of the LTS flowsheet over the existing flowsheet are (1) higher filtration rate (8.5 gpm vs.5.0 gpm), (2) less frequent filter cleaning/washing cycles (once every 200 batches vs. once every 20 batches), and (3) more reliable systems with higher sustainable attainment (70% attainment vs. 50% attainment).

COMPARISON OF SALT PROCESSING THROUGHPUTS BEFORE AND AFTER LTS FLOWSHEET

Table 3 compares of salt processing throughputs before and after LTS flowsheet. As a result, the projected salt processing rate will increase from 1.1 Mgal/yr fresh waste processing rate to about 2.4 Mgal/yr fresh waste processing rate; which represents 118% increase in salt annual processing rate. With LTS flowsheet, the process bottleneck is also shifting from the ARP/MCU system toward the Feed Prep system. Frequent filter cleanings & washings might create accumulation of sodium oxalate issues in down-stream process vessels. Sodium oxalate is snowy white particles which had caused significant process upset. Case in point, in 2014, a significant amount of sodium oxalate accumulated inside the Salt Solution Receipt Tank (SSRT) which was agitated unintentionally resulting in an MCU extended outage of ~2,420 hours.

Table 3: Process Parameters Comparison for Current ARP/MCU flowsheet and LTS Flowsheet

	Existing Flowsheet	LTS Flowsheet
# Blend Tank	1 Blend Tank MST strike in 96-H fresh waste batch size = 3,715 gal	1 Blend tank MST strike in Tank 21 fresh waste batch size = 1 Mgal
512-S Filtration Rate (gpm)	5	8.5
CSSX Processing Rate (gpm)	8.5	8.5
Filter Cleaning once every X batches	X = 20	X = 200

OTHER POTENTIAL BOTTLENECKS AND LIFE CYCLE COST SAVING

A total of thirteen bottlenecks (improvement opportunities) were identified and ranked based on impacts of removing a bottleneck on salt waste processing throughputs. The "Large Tank Strike" would address two of the top three bottlenecks. The # 2 improvement opportunity is to have a second one million gallon blend tank. The improvement opportunity would increase salt processing throughput by 41%, with a cost of ~$100 million. According to the modeling results, if all identified bottlenecks can be eliminated, the ARP/MCU will be able to reach 4 Mgal/yr annual throughput rate vs. the current annual rate of ~1 Mgal/yr. Impact on The Return-On-Investment (ROI) for some bottlenecks is too small to justify tying up limited resources. The model clearly provides a priority list for deployment of limited resources based on quantified ROIs to save cost and expedite tank closures schedule. With just LTS improvement alone, the projected Salt Processing throughput is expected to increase from ~1 Mgal/yr to ~2 Mgal/yr. SRR is expected to receive estimated ~$400 million EM funding, subject to funds availability, for FY2015 budget. By doubling salt processing throughput rate, it is estimated the site tank closure date would be shorten by ~1–2 years. With the above assumptions, the life cycle cost saving of the LTS would be ~ $400–$800 million.

REFERENCES

(1) Henley, E.A. & Yeung M.R., 2014. "Tank 21 Solid Particle Settling Rate Calculation", Calculation Number S-CLC-H-01268. Location: Savannah River Site, SC.
(2) Lee, E.D., "Insoluble Solids Settling in Tank 51 During Baseline Runs", WSRC-RP-96-87, Rev. 0, May 1996.
(3) Staub, A.V., Technical Task Request for Tank 21 Solid Particle Settling Rate Calculation, X-TTR-H-00053, Rev. 0, November, 2014.
(4) Baron, P.A., and Willeke, K., Aerosol Measurement, Second Edition, John Wiley and Sons, 2001.
(5) De. Nevers, N., Air Pollution Control Engineering, McGraw-Hill International Editions, 1995.
(6) Staub, A.V., "MST Particle Size Distribution Data for Use in Parge Tank Strike Settling Time Determination." X-ESR-H-00720, Rev. 0, October 2014.
(7) FFA Appendix L, "Statement of Resolution of Dispute Concerning Extension of Closure Dates for Savannah River Site High-Level Radioactive Waste Tanks 19 and 18 effective in November 2007, and amendments"
(8) Ledbetter, J., "Input Data and Assumptions for the Concentration, Storage, and Transfer Facilities (U)," S-CLC-G-00235, Rev. 7, July 2011.
(9) Chew, D.P. & Hamm, B.A., SRR-LWP-2009-0001, Rev. 19, May 2014.
(10) Le, T.A., Salt Batch Plan in Support of System Plan R-19, SRR-LWP-2014-00003, 2014.

427760 Prediction of Radiolytic Benzene Generation in a Waste Tank at the Savannah River Site

Abstract:

Back in the 1990s, an in-tank precipitation process at the Savannah River Site (SRS) was designed to remove radioactive cesium and strontium from aqueous high-activity waste. The process begins when saltcake stored in the SRS Tank Farms is dissolved in water and transferred to the processing tank. Dilution water is added to adjust the salt concentration. Sodium tetraphenylborate and sodium titanate slurry are added to the salt solution. Cs-137 is removed by precipitation with sodium tetraphenylborate, and Sr-90 is removed by adsorption on sodium titanate particles. The tank dilute slurry is concentrated by filtration. The resulting decontaminated filtrate salt solution is processed as a low-activity waste. The concentrated precipitate remaining in Tank 48 is ultimately processed as high-activity waste in the Defense Waste Processing Facility (DWPF) vitrification plant.

The radiolytic generation of benzene in soluble and solid tetraphenylborate (TPB) compounds raised safety concerns. Benzene, when released from solution, may exceed the lower flammable limit if adequate purge air flow through the tank gas space is not maintained. Benzene reaches the vapor space by the following mechanisms: (1) formation of dissolved ("free") benzene by radiolysis of slurries containing TPB compounds; (2) formation of "trapped" benzene by radiolysis of solid TPB compounds; (3) release of "trapped" benzene during dissolution of the solids; (4) adsorption of "free" benzene on the surface of the solids; and (5) evaporation of benzene into the gas space. To predict benzene emissions, a FORTRAN dynamic model was first developed, and later converted into Aspen Custom Modeler (ACM). Due to safety issues, the in-tank precipitation process was replaced by other processes (i.e., Actinide Removal Process (ARP) to remove strontium, Modular Caustic Side Cesium Extraction Unit (MCU) to remove cesium). The processing tank has remained idle since 2000.

This paper discusses the model development and presents the computational results for the designed in-tank precipitation process and for the current processing tank waste content.

427536 Pilot-Scale Filter Testing of a Process to Treat Radioactive Liquid Waste

Abstract:

P

The Savannah River Site is developing processes to treat radioactive liquid waste. One of these processes is the actinide removal process. This process adds monosodium titanate (MST) to the waste to sorb strontium and select actinides. The MST and entrained sludge (i.e., insoluble metal oxides) are removed from the waste by a crossflow filter. Since start up, filter fouling has limited the throughput of the process.
The authors constructed a pilot-scale crossflow filtration apparatus to perform tests to understand the causes of the filter fouling and to improve the filter flux. This presentation will discuss the construction of the pilot-scale filter and the filter testing results.

435504 Preparation of Geopolymers from Wastes

Abstract:

Geopolymers, an inorganic polymer composed of alumina and silica network with alkali metal ions, were prepared from wastes, such as incinerated ash of sewage sludge or fly ash from power plants. We have examined the effect of the initial compositions of the raw materials controlling the ratio of alumina, silica, and alkali in addition to the solid / liquid (water) ratios on the polymerization reactions of the geopolymer. In some cases, meta-kaolin and silica powder were added to adjust the target composition. We demonstrated the region where geopolymer is formed in a ternary diagram of which the three axes are fraction of silica, alumina, and alkali under a fixed solid /water ratio. The formation of geopolymers was confirmed by checking the hardening of the mixtures, and the solid product was characterized with XRD and BET measurements in addition to the strength tests. The surface area and the pore size distribution of the geopolymers greatly depended on the initial composition of the raw materials. In some cases, zeolites are formed: under the ambient temperature and pressure conditions. The surface area varied in the range of about 8.6 m^2/g to 460 m^2/g. We have found the surface areas of the geopolymers can be well correlated with the two ratios of silica / alkali and alumina / alkali.

Savannah River National Laboratory™

OPERATED BY SAVANNAH RIVER NUCLEAR SOLUTIONS

We put science to work.™

Technical Bases for the DWPF Melter Off-gas Flammability Control under the Glycolic-Acid Flowsheet

A. S. Choi
F. C. Johnson
J. R. Zamecnik

AIChE Annual Meeting
November 8-13, 2015

Savannah River NUCLEAR SOLUTIONS
FLUOR • NEWPORT NEWS NUCLEAR • HONEYWELL

Agenda

- DWPF melter off-gas flammability control strategy
- Overview of models
- Pilot melter runs with new flowsheet feeds
- Revision of models
- Conclusions and path forward

Defense Waste Processing Facility (DWPF)

- Began immobilizing high-level waste (HLW) in a borosilicate glass matrix in March 1996 and has since produced 4,000+ canisters of glass.
- Feed contains organic carbon species (formate, antifoam, strip effluent, oxalate and coal)
  - → potential flammability
  - → DSA
- Slurry feeding
  - → off-gas surges
  - → spike in H_2/CO

Defense Waste Processing Facility (DWPF)

- Began immobilizing high-level waste (HLW) in a borosilicate glass matrix in March 1996 and has since produced 4,000+ canisters of glass.
- Feed contains organic carbon species (formate, antifoam, strip effluent, oxalate and coal)
  - → potential flammability
  - → DSA
- Slurry feeding
  - → off-gas surges
  - → spike in H_2/CO

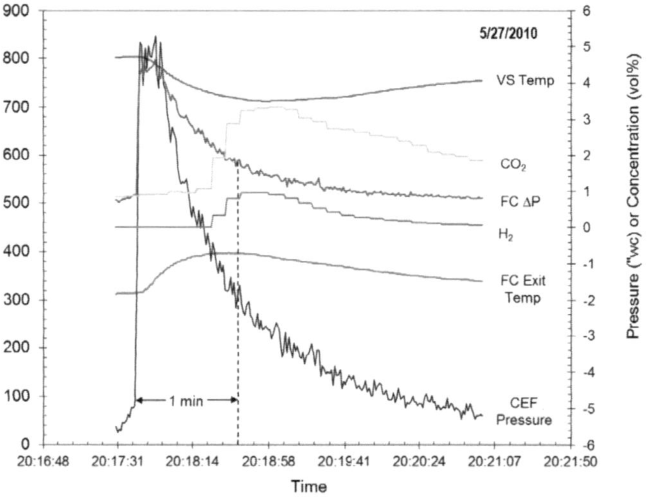

DWPF Melter Off-Gas Flammability Control Strategy

- **Mitigation of potential flammability by off-gas monitoring is not an option.**
  - Difficult to maintain safety-class instruments in a radioactive, particulate-laden system
  - Fast transient with duration shorter than instrument lag times
- **Describe time-dependent off-gas flammability potential as a function of process variables using computer models.**
- **Calibrate and validate computer models using pilot melter data.**
- **Run computer models to define the operating window for all process variables affecting flammability (TOC, T, air purges, etc.) under transient conditions.**
- **Perform uncertainty analysis on theoretical limits to derive actual field limits.**
  - Technical Safety Requirements (TSR) on TOC
  - Feed interlocks on temperature and air purges

DWPF Melter and Primary Off-Gas System

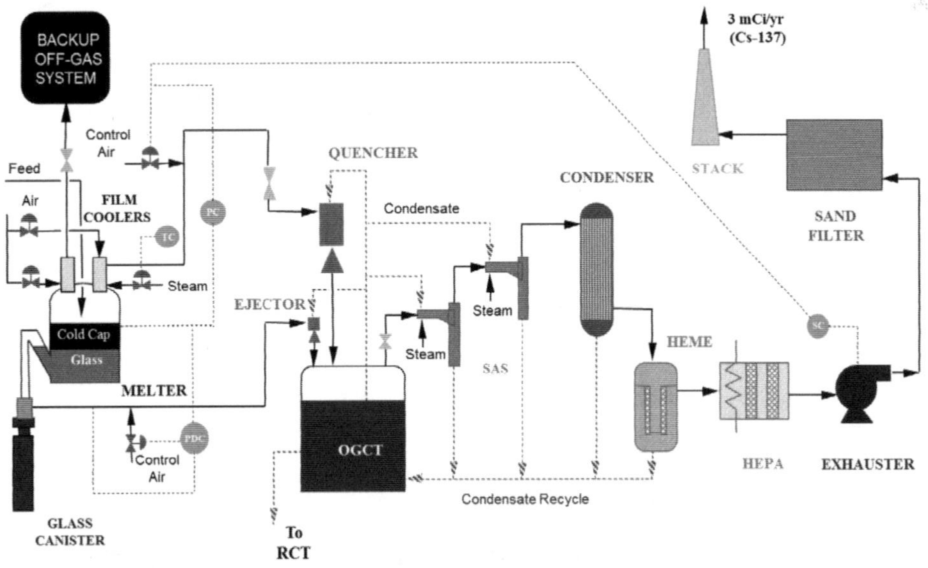

4-Stage Cold Cap Model

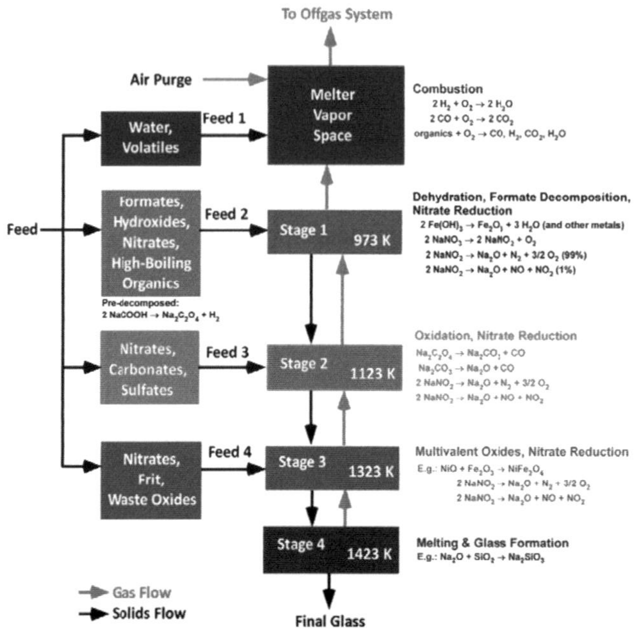

DWPF MOG Flammability Model

4-Stage Cold Cap Model

- A multi-stage, countercurrent equilibrium reactor.
- Gas and solid products reach thermodynamic equilibrium in each stage.
- Accounts for strong solution non-ideality present in waste glass melt.
- Relies upon phase diagrams and engineering judgement to identify stable species and multiple solutions.
- Output: glass composition (Fe^{2+}/Fe^{total}) and calcine gas flows (H_2, CO, CO_2. NO_x, etc.)

Melter Off-Gas Dynamics Model

- 5-component transient heat/mass balance across 40+ unit operations.
  - Input streams: volatile organics in feed and calcine gases from the cold cap
- Includes 2-step global combustion of volatile organics and oxidation of CO and H_2.
- Simulates all major primary & backup off-gas system hardware and DCS software logic including 22 control loops and 26 dynamic valves.
- Successfully simulated: startup & shutdown, off-gas surges, switchover & switchback, various equipment failures, feed tube flushing, etc.

Calculated vs. Measured Concentrations of H_2 and CO in the 1/80 Scale Melter Off-Gas

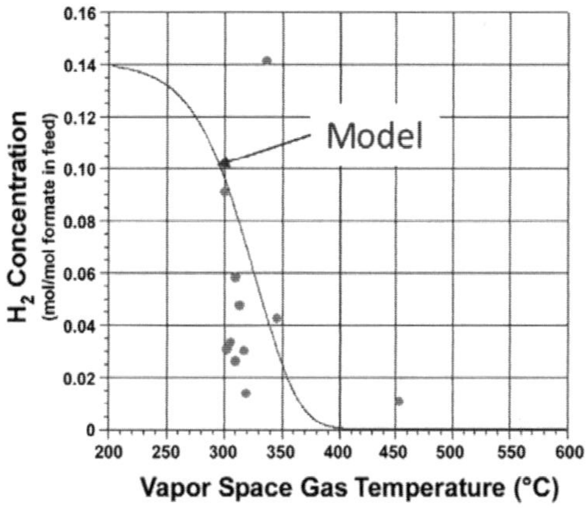

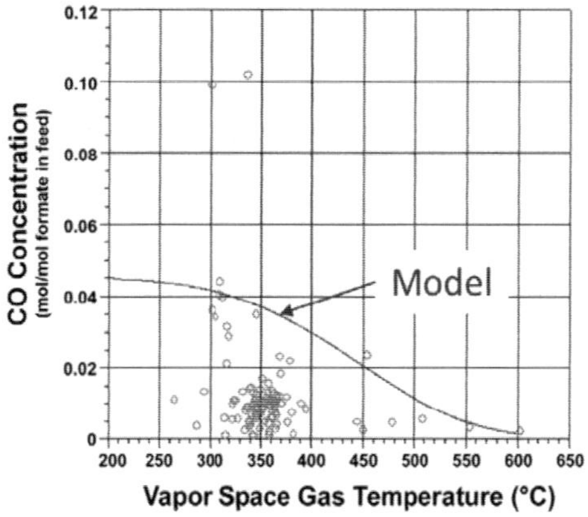

Recent DWPF Melter Surge vs. Model

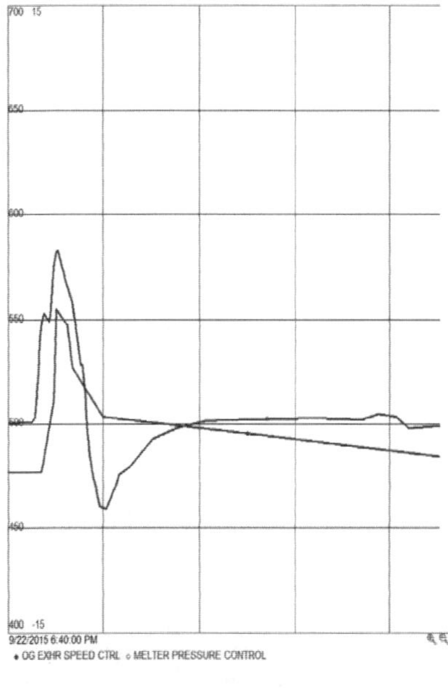

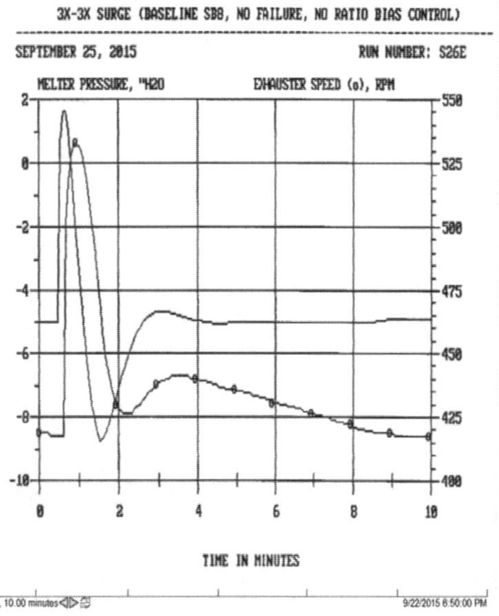

Current DWPF Controls for Bubbled Operation

- ***Maximum TOC** = $a + b\,[NO_3] + c\,[NO_3]^2 + d\,[NO_3]^3 + e\,[NO_3]^4$*

  $10{,}000 \leq [NO_3] \leq 70{,}000$ mg/kg slurry

- **Maximum antifoam = 728, 894 and 1,017 gallons per batch**
- **Minimum melter vapor space temperature (TI4085D) = 460 °C**
- **Minimum total melter air purge (FIC3221A) = 900 lb/hr**
- **Minimum backup film cooler air purge (FIC3221B) = 233 lb/hr**
- **Maximum feed rate = 1.5 GPM (at 45 wt% total solids)**
- *Bounding off-gas surge = 9X normal condensable and 5X normal non-condensable flows*

9X/5X Surge Simulation of a Bounding SB8 Feed at 25,000 ppm NO_3

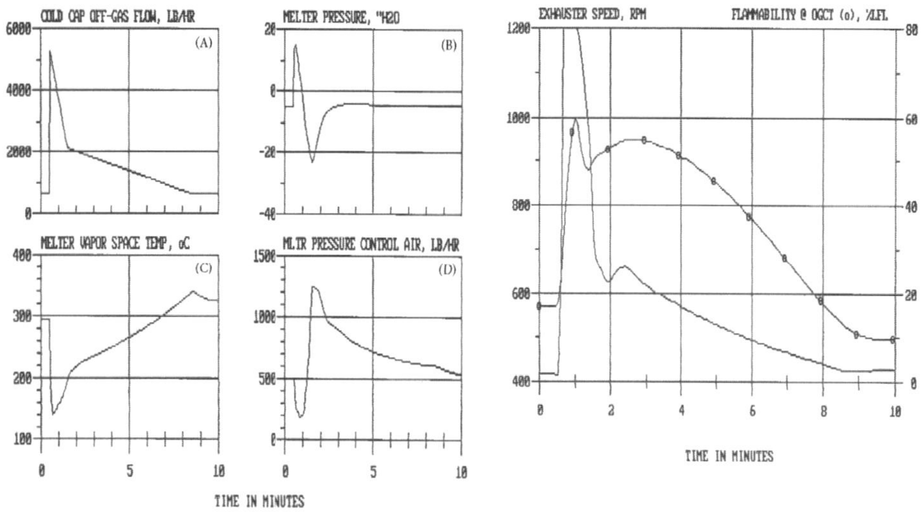

Formic vs. Glycolic Acid

- Reduction: $MnO_2 + 2\,HCOOH \rightarrow Mn(COOH)_2 + H_2O + \frac{1}{2}\,O_2$

  $HgO + HCOOH \rightarrow Hg + CO_2 + H_2O$

- *Catalytic decomposition:* $HCOOH \rightarrow CO_2 + H_2$

Formic vs. Glycolic Acid

- Reduction: $MnO_2 + 2\,HCOOH \rightarrow Mn(COOH)_2 + H_2O + \frac{1}{2}\,O_2$

  $HgO + HCOOH \rightarrow Hg + CO_2 + H_2O$

- *Catalytic decomposition:* $HCOOH \rightarrow CO_2 + H_2$

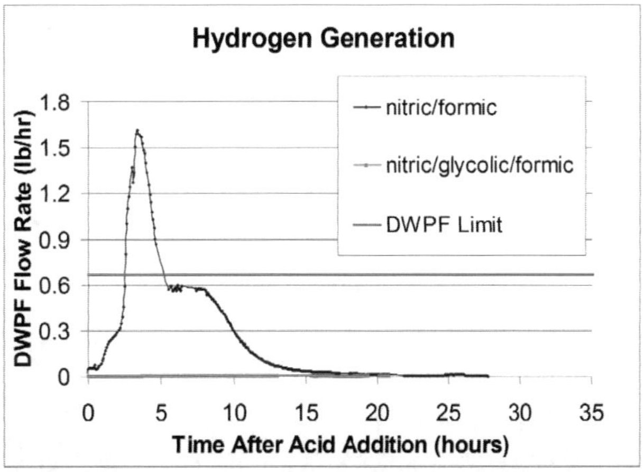

CEF

- 30 inch tall, 20 inch diameter Inconel 690 vessel
- 1/12th scale of DWPF melter
- Vapor space pressure at -5" H_2O
- 6 separate heating zones (resistance)
- 2 bubblers (Ar)
- 2 cameras for cold cap viewing
- Active air purge

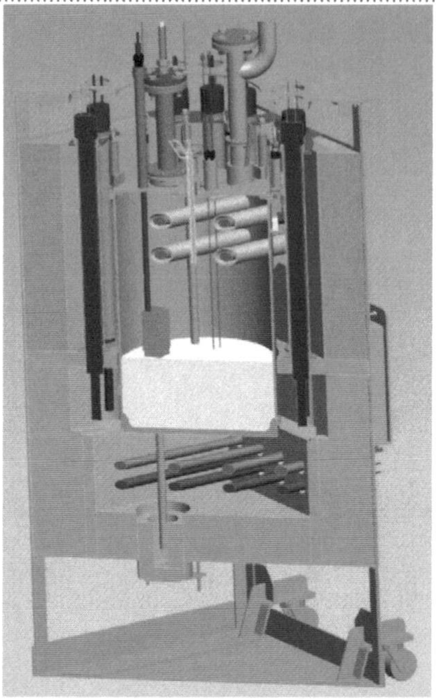

Feed

- SB6 simulant with no noble metals or Hg
- Nitric/glycolic acids added at 100% and 125% of the stoichiometric amount
- Further remediated for REDOX control
- Frit 418 added at 36% target waste loading
- Diluted to either 45% or 42% total solids

CEF Off-Gas System and Instrumentation

- Consists of a film cooler, scrubber, off-gas condensate tank and blower
- Film cooler (air injected tangentially through slots)
- Venturi scrubber with re-circulated condensate
- Prototypic configuration of the film cooler, scrubber and interconnecting piping
- Mass spectrometer
  - *H_2, O_2, N_2, NO, Ar, CO_2 and SO_2*
  - *Sampling interval: 6-7 seconds*
- Fourier transform infrared (FTIR) spectrometer
  - *CO, CO_2, H_2O, N_2O, NO, NO_2, SO_2 and CH_4*
  - *Sampling interval: 3.5 seconds*
- Gas chromatograph
  - *H_2, O_2, N_2, CO_2 and N_2O*
  - *Sampling interval: 4 minutes*

Feed Analysis at 45 wt% Total Solids – Metals/Frit

Oxide (wt%)	Average Measured Melter Feed Composition	Average Calculated Glass Composition	Average Measured Glass Composition
Al_2O_3	9.98	9.49	9.21
B_2O_3	4.90	5.12	4.86
CaO	0.58	0.59	0.59
Fe_2O_3	11.45	11.04	10.77
Li_2O	4.85	5.12	4.95
MgO	0.54	0.50	0.58
MnO_2	3.94	3.88	3.90
Na_2O	12.07	11.65	11.57
NiO	1.34	1.34	1.35
SiO_2	49.15	49.77	49.67
Total	98.80	98.50	97.45

Feed Analysis at 45 wt% Total Solids - Anions

	100% Acid (mg/kg)	125% Acid (mg/kg)
NO_3	58,765	66,774
$C_2H_3O_3$	34,494	42,765
- $C_2H_3O_3$ (free)	7,793	12,606
TOC	13,322	15,573
- formate	427	159
- glycolate	9,987	7,730
- oxalate	337	325
- free formic	40	39
- free glycolic	1,055	5,959
- antifoam	1,476	1,360
pH	5.03	3.21

CEF Run Sequence

	Duration (day)
Melter Turnover	1
Steady State @ 700, 600, 500, 400, 350, <300 °C	11
- 125%, Bubbled	
- 125%, Non-bubbled	
- 100%, Bubbled	
- 100%, Non-bubbled	
Surge Testing @ 700 °C	9
- 100%, Bubbled	
- 100%, Non-bubbled	
Antifoam Spike @ 700, 600, 500, 400, 350, <300 °C	3
- 100%, Bubbled	
Drain & Shutdown	1
Total	25

H_2 Spikes

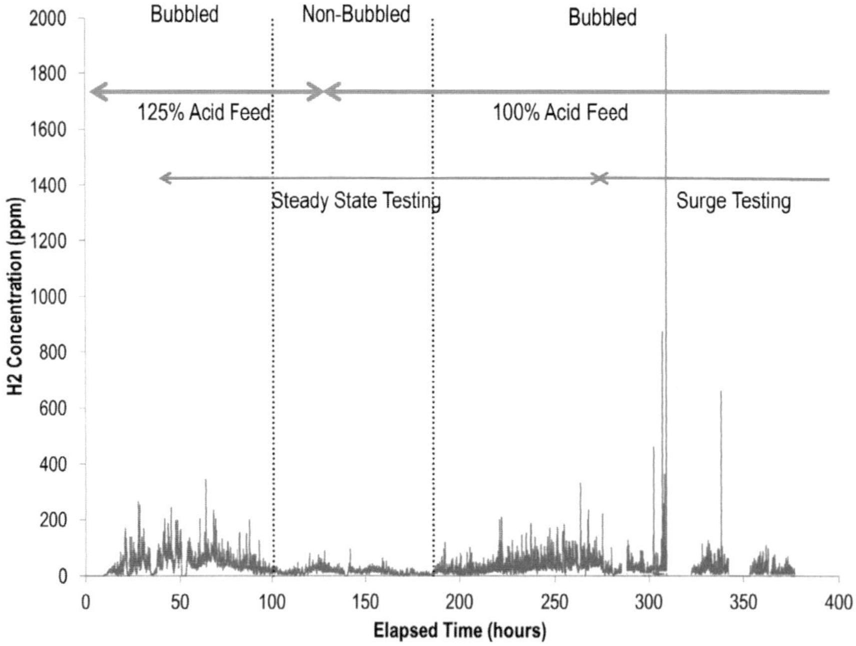

Bubbled Surge Testing (100% stoichiometry feed, 45 wt% total solids)

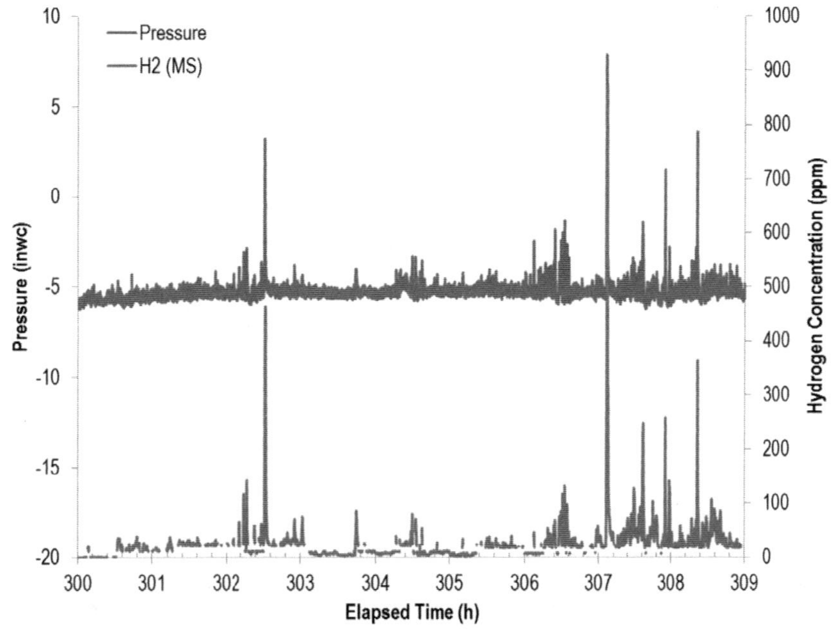

Comparison of Pressure Spikes, Formic vs. Glycolic Acid Flowsheet

Pressure Spike above Baseline ("H_2O)	Bubbled		Non-bubbled	
	Formic 2010	Glycolic 2014	Formic 2010	Glycolic 2014
1 - 2	98	1,164	28	1
2 - 3	42	43	15	0
3 - 4	27	15	3	0
4 - 5	7	6	1	0
> 5	18	14	0	1
Total	192	1,242	47	2
Test Duration (day)	4.3	4	2.5	1
Frequency (#/day)	45	311	19	2

13” H_2O Pressure Spike during Phase 2 Bubbled Test at 700 °C

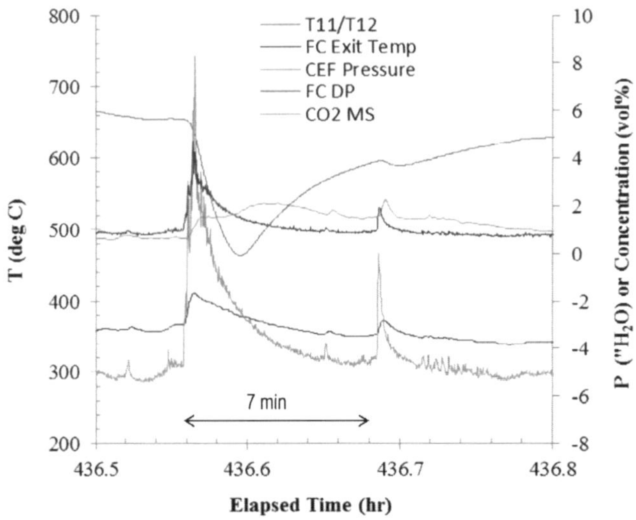

Surge Testing (~7” H_2O pressure spike)

Average CEF Operating and Off-Gas Data during Steady State Runs

SS Run	VS Temp (°C)	Feed Rate (g/min)	FC Air (scfm)	Purge Air (scfm)	FC Exit T (°C)	Melter P ("H2O)	FC Exit P ("H2O)	Glass Temp (°C)	H_2 (MS) (ppm)	CO (FTIR) (ppm)	CO_2 (MS) (vol %)	NO (MS) (ppm)	NO_2 (MS) (ppm)	N_2O (FTIR) (ppm)
125%_B_1X	705	221	15.64	0.65	366	-5.22	-5.97	1,033	43	224	0.919	2,183	1,430	391
	625	161	15.92	0.68	299	-5.35	-5.99	1,032	70	180	0.676	1,898	981	266
	481	114	15.88	15.12	256	-4.90	-6.20	1,025	42	32	0.312	917	388	106
	415	111	16.08	22.28	234	-4.74	-6.42	1,029	28	23	0.245	793	364	90
	360	93	8.54	40.14	244	-3.48	-5.91	1,011	18	22	0.183	498	228	68
	270	72	6.20	50.68	203	-1.16	-3.85	1,035	19	19	0.162	441	184	59
125%_NB_1X	709	100	15.73	0.45	325	-5.47	-6.02	1,083	13	119	0.431	1,360	607	219
	393	51	15.73	14.98	212	-5.46	-6.56	1,085	23	14	0.189	517	275	75
	351	44	15.93	20.89	200	-5.04	-6.43	1,083	16	11	0.157	374	279	56
	302	35	15.91	29.62	183	-4.96	-6.81	1,084	12	6	0.119	332	213	32
100%_B_1X	705	170	15.25	0.52	358	-4.81	-5.50	1,050	19	161	0.659	1,947	1,091	220
	592	146	15.32	0.51	285	-4.66	-5.25	1,042	37	109	0.512	1,682	866	165
	471	123	15.32	6.09	235	-4.42	-5.22	1,041	31	55	0.399	1,303	673	115
	373	102	15.50	25.94	219	-5.06	-6.96	1,025	18	20	0.218	689	340	56
	323	88	6.12	45.90	224	-1.13	-3.61	1,031	13	18	0.130	545	268	674
100%_NB_1X	697	125	16.14	0.32	323	-5.27	-5.77	1,086	17	117	0.476	1,445	671	152
	600	99	16.55	0.32	268	-5.70	-6.15	1,085	23	85	0.355	1,137	518	125
	496	84	16.80	0.33	214	-6.00	-6.42	1,068	19	26	0.297	993	455	96
	410	56	16.83	13.22	209	-5.46	-6.49	1,074	7	8	0.121	502	244	634
	344	48	16.10	27.93	201	-4.46	-6.28	1,078	6	4	0.094	307	196	364
	326	42	9.67	34.77	217	-4.02	-5.99	1,079	6	4	0.095	284	252	348
100%_NB_2X	722	103	16.14	0.19	334	-2.92	-3.67	1,097	6	154	0.577	0.527	0.590	1,522
	604	47	15.97	0.41	292	-5.38	-6.14	1,071	60	213	0.484	0.457	0.493	1,300
	519	57	15.99	0.42	222	-5.12	-5.77	1,066	10	89	0.256	0.252	0.267	787
	397	49	15.85	19.00	211	-4.63	-6.29	1,052	9	51	0.155	0.159	0.164	392
	323	39	10.01	27.97	211	-4.04	-5.97	1,067	9	22	0.118	0.131	0.126	287
	293	37	5.69	34.39	227	-2.80	-4.88	1,066	10	27	0.110	0.119	0.118	243

Savannah River National Laboratory
OPERATED BY SAVANNAH RIVER NUCLEAR SOLUTIONS
We put science to work.
25

Cumulative H_2 Evolution during Batch-Melting in Ar-Purged Crucible

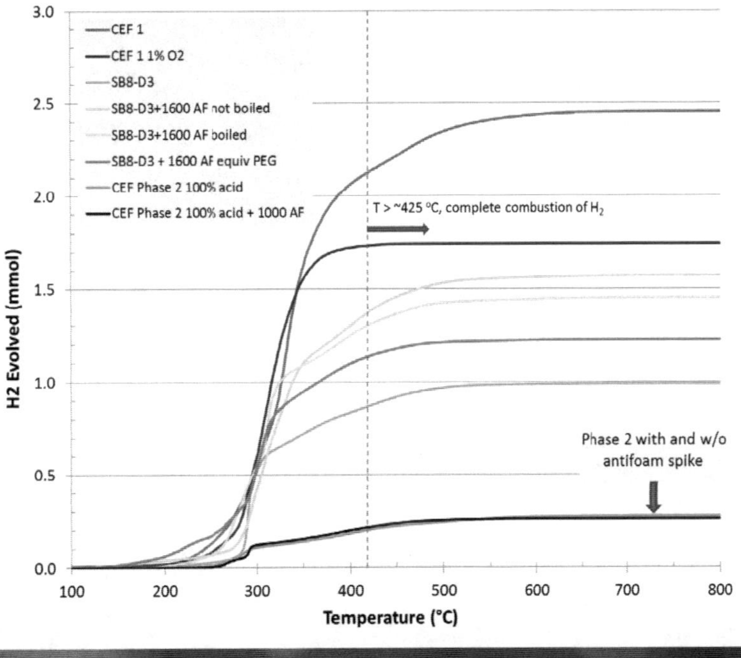

Savannah River National Laboratory
OPERATED BY SAVANNAH RIVER NUCLEAR SOLUTIONS
We put science to work.
26

Comparison of H_2 Generation from Formic vs. Glycolic Acid Flowsheet Feeds – CEF Data

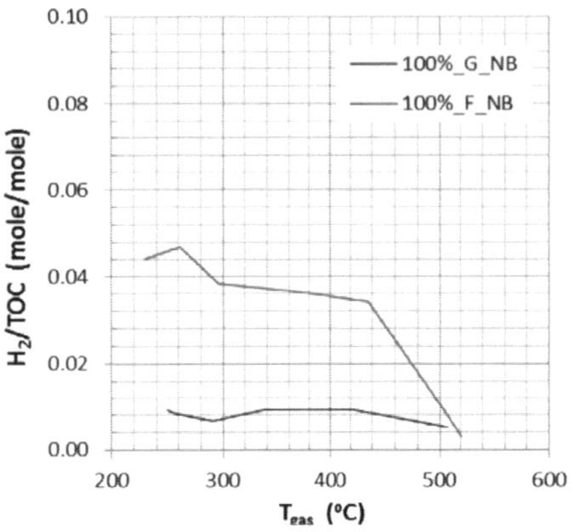

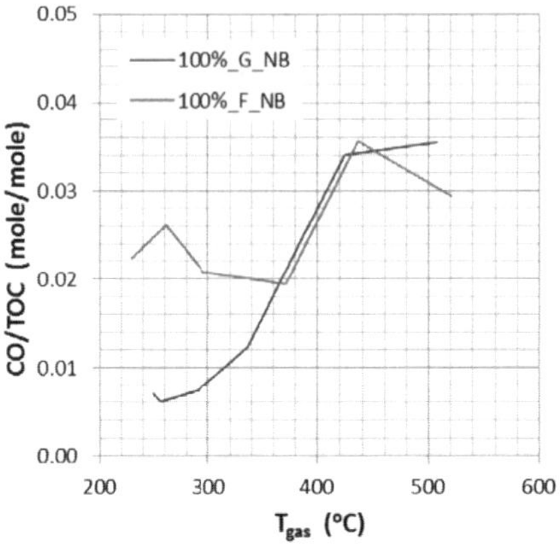

Revision of Existing Models

- **Fractional removal of nitrate from Cold Cap model input**
  - 4X higher nitrate forces the model to predict no H_2 in the calcine gases vs. <100 ppm measured and spiked to 1,000 ppm → 22% (100% stoichiometry) and 29% (125% acid stoichiometry)
- **Free glycolic acid decomposition:** $C_2H_4O_2 \rightarrow 2CO + H_2 + H_2O$

	100%_NB	100%_B	125%_NB	125%_B
$k_{o,G}$ (1/sec)	2,246.44	238.22	361.37	24.36
$E_{a,G}$ (Kcal/gmole)	13.05	9.40	10.76	7.70
$E_{a,G}$ (Btu/lbmole)	23,486	16,924	19,375	13,854

- **Oxidative decomposition of remaining glycolic acid:**

$$C_2H_4O_3 + \frac{3}{2}O_2 \rightarrow 2CO_2 + 2H_2O$$

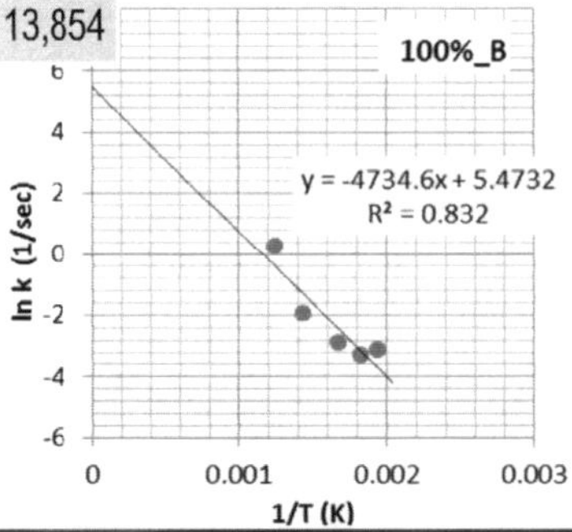

Revision of Existing Models (Cont'd)

- **Oxidation of H_2 and CO modeled using global 1st order kinetics**

	100%_NB	100%_B	125%_NB	125%_B	Current
$k_{o,H2}$ (1/sec)	1,770.47	1,952.37	36,534.05	67.66	2.795 E7
$E_{a,H2}$ (Kcal/gmole)	10.01	10.02	13.65	5.64	21.63
$E_{a,H2}$ (Btu/lbmole)	18,015	18,032	24,562	10,158	38,940

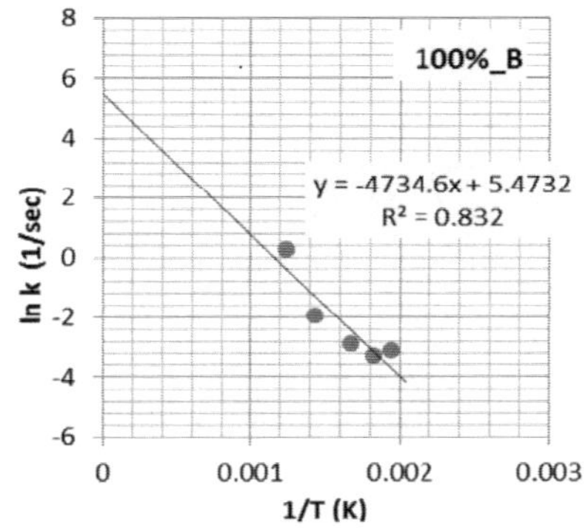

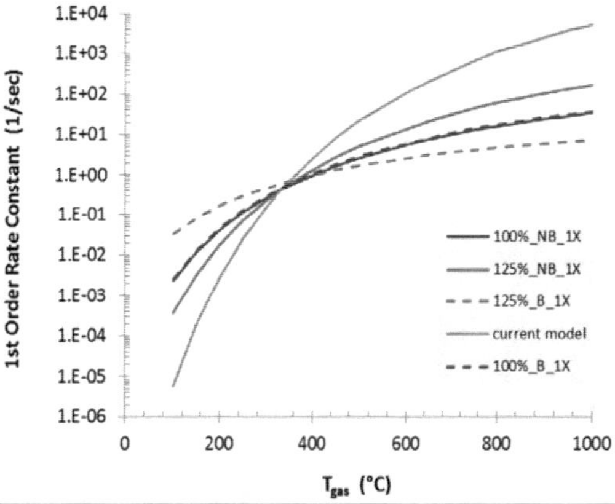

Limited Testing of Revised Kinetic Models

- **+13" H_2O CEF Pressure Spike Data:**

Operating Mode	Variable	Data	Model
Steady State	CO (ppm)	188	235
	H_2 GC (ppm)	60	51
	H_2 MS (ppm)	46	
Bounding	CO (ppm)	694	745
	H_2 GC (ppm)	215	488

Conclusions and Path Forward

- On a per carbon basis, the glycolic acid flowsheet feed produces ¼ of the H_2 produced by the formic acid flowsheet feed (in the temperature range relevant to the DWPF melter off-gas flammability safety basis).
- A new off-gas surge basis has been defined for the glycolic acid flowsheet; it consists of 12X normal condensable and 4X normal non-condensable flows (12X/4X) for bubbled operation vs. current 9X/5X.
- 1/80th scale Slurry Melt Rate Furnace (SMRF) is being prepped to produce data using different feeds to support full model validation.
- Once validated, the models will be used to generate a new set of TSR and feed interlock limits for DWPF bubbled and non-bubbled operations.

Acknowledgements

- **Work was performed for the U.S. Department of Energy under the contract number DE-AC09-08SR22470.**
- **Funding was provided by Savannah River Remediation (SRR).**
- **The Environmental and Chemical Processing Technology (E&CPT) staff at SRNL provided 24/7 run coverage during the 25-day CEF test.**
- **Feed, condensate and glass samples were analyzed by AD and PSAL at SRNL.**

422515 A Better Understanding for Cold-Cap Reactions of Nuclear Waste Feeds through Quantitative Evolved Gas Analysis

Abstract:

In the vitrification of nuclear wastes, waste slurry mixed with glass former chemicals (melter feed) is charged into an electrical glass-melting furnace (a melter) where it becomes a cold-cap, a layer of reacting melter feed floating on a pool of molten glass. Within the cold cap, the feed experiences multiple gas-evolving reactions. Understanding these reactions is critical because the reaction enthalpy and the generation of foam significantly influence the mass and heat transfer within the cold-cap, thus affecting the melting rate. A kinetic model based on thermogravimetric analysis (TGA) provides invaluable insight into the cold-cap reactions, but it does not identify their chemistry. We employed the thermogravimetry-gas chromatography–mass spectrometry (TGA-GC–MS) combination to perform qualitative and quantitative evolved gas analysis (EGA) for the cold-cap reactions. By adding the 'chemical identity' to the TGA-based kinetic model, the EGA allows us to recognize possible reactions/reactants over the temperature range of the feed-to-glass conversion. The quantitative analysis was achieved by relating the weighted sum of intensities of individual gases in linear proportion with the differential thermogravimetry. The proportionality coefficients were obtained by three methods based on the stoichiometry, least squares, and calibration. The linearity was shown to be a good first-order approximation, in spite of the complicated overlapping reactions.

AUTHOR INDEX

AUTHOR INDEX